Social Computing with Artificial Intelligence

Xun Liang

Social Computing
with Artificial Intelligence

清華大學出版社
TSINGHUA UNIVERSITY PRESS

Springer

Xun Liang
School of Information
Renmin University of China
Beijing, China

ISBN 978-981-15-7762-8   ISBN 978-981-15-7760-4   (eBook)
https://doi.org/10.1007/978-981-15-7760-4

Jointly published with Tsinghua University Press
The print edition is not for sale in China (Mainland). Customers from China (Mainland) please order the print book from: Tsinghua University Press, Beijing, China.

This Springer imprint is published by the registered company Springer Nature Singapore Pte Ltd.
The registered company address is: 152 Beach Road, #21-01/04 Gateway East, Singapore 189721, Singapore

Preface

Over recent years, the social network platform has seen unprecedented rapid development. These network platforms (such as Facebook, Twitter, Sina Weibo, Wechat, Netease News) not only gather a large number of users, but also provide huge potential user resources. These user resources promote the production of massive data sets in the field of business and scientific research, which are of vital significance for analyzing social group behavior and exploring potential huge social value. Therefore, a new discipline emerged at the historic moment—social computing.

As a new interdisciplinary research field, social computing, there is no recognized definition at present. However, we can analyze concepts from the background of social computing and generalize social computing as "using social methods to calculate society." The so-called socialization method is a method that centers on "grass-roots" users and relies on "grass-roots" users. It is a method of synergy and swarm intelligence, a mode of thinking from individual to whole, from micro to macro. Many incidents have been developed into a major social event by countless netizens' words and insignificant microbehavior. From this point of view, social computing is a computational model of swarm intelligence. It can be seen that the essence of social computing is the process of intelligent analysis of large social network data.

The research object of social computing is society, including physical society and virtual network society. The former mainly refers to our traditional sense of society, such as a country or region; the latter mainly refers to the web-based virtual network community. Virtual network society is a reflection of real physical society. The ultimate goal of studying virtual network society is to serve the management of real physical society. The study of social network activities will help to find the stable network that has remained for a long time, excavate the huge potential value, and provide important help for decision-makers to make more effective decision analysis.

The study of social network cannot be separated from the collision with artificial intelligence or machine learning, which is the most active branch of computer science at present. In this book, we also introduce supervised, unsupervised, and semi-supervised learning models as well as more specific state-of-the-art artificial intelligence algorithms, such as deep learning, reinforcement learning, brother learning, and epiphany learning. As a field leading the revolutionary change of information technology, the rise of artificial intelligence has promoted the rapid development of the Internet and the wide application of social networks, making the dissemination of information faster and the fermentation of public opinion warmer. Therefore, public opinion also needs modern means to manage. Public opinion is a kind of public composed of individuals and various social groups. In a certain historical stage and social space, it is the sum of various emotions, willingness, attitudes, and opinions held by various public firms which are closely related to their own interests. Network is the carrier of public opinion dissemination in modern society and has given new characteristics to the dissemination of public opinion.

Social network public opinion is a specific form of social public opinion, and a collection of common opinions which have certain influence and tendencies on certain phenomena, problems, or specific things publicly expressed by the public on the social network. However, the social network public opinion further narrows the network public opinion, which is limited to the relevant public opinion content of this subject. In recent years, public opinion monitoring has become a necessary technology for the state to manage the Internet. Using computer intelligence technology to transform various human emotions into real numerical data has become a research hotspot.

This book consists of twelve chapters, starting with low-level data processing, and gradually transiting to higher-level social computing and its applications. It is divided into three parts: Part I, data (including Chaps. 2 and 3), introduces the social computing data collection and data analysis algorithm, which correspond to social computing in the four areas of social data perception, i.e., knowledge discovery, individual and group community modeling, information dissemination and network dynamic evolution. Part II, models (including Chaps. 4–8), introduces the social computing model in three aspects: online crowd opinion content mining, community network structure, dynamic information dissemination, network propagation mechanism, which correspond to the areas of social computing. Part III, applications (including Chaps. 9–12), introduces the application cases of decision-making in the fields of social computing to support social practical problems.

In order to help readers have a stronger appreciation of the concepts, the book contains a large number of data, charts and references, including data and algorithms that we think are vital, available and explanatory, so that readers can dig into social computing through this book as much as possible.

This work was supported by the National Social Science Foundation of China (18ZDA309) and the National Natural Science Foundation of China (71531012). Thanks also go to the author's colleagues and students in preparing, handling and organizing the data and documents, including Shimin Wang, Hua Shen, Jin Ruan, Shusen Zhang, Yang Xue, Xuan Zhang, Xiaoping Zhou, Yuefeng Ma, Jinshan Qi, Bo Wu, Mengdi Liu, etc.

Beijing, China   Xun Liang

Contents

Chapter 1
Introduction

1.1 Research Background

To date, people live life in online social network. We check our emails regularly and post messages on social media networks like Twitter, Facebook, etc. Our behaviors may bring about enormous data that hold the characteristics as 4Vs: volume, variety, velocity, and value. These data are of great importance for analyzing social group behavior and exploring potential huge social value. Therefore, in recent years, a new discipline emerged as the times require. The social computing is a field that leverages the ability to collect and analyze data at a scale that takes use of the potential online crowd wisdom. This is a data-driven interdisciplinary subject involving information science, behavioral science, psychology, systems engineering and decision science, which could provide various solutions to complex social problems [1].

With the outbreak of the big data online, social computing produced a variety of data acquisition modes, such as open data sharing in research communities, interactive data recording through video, people's links with email data. In addition, trajectory data, electronic communication data, online data, etc., which are all related to people's social behavior. Besides, social computing makes it possible to solve many traditional sociological problems efficiently. For example, complex problems can be greatly simplified by establishing system model and designing corresponding big data algorithm. According to the purpose of decision-making, a variety of schemes are designed and compared to provide the basis for the implementation of decision-making optimization. Beyond the space-time limit of the research object, the simulation of the research object is realized on the computer, and the tracking research or prediction research of the object which is difficult to be realized by other ways is completed. It is based on the study of objective factors and their interrelations, as well as the rigorous and accurate operation of computers. Therefore, it has considerable objectivity. In addition, a large number of advanced data analysis technologies emerged in the field of social computing.Big data analytics is the application of advanced analytics tools to big data sets. These analysis tools include prediction analysis, data mining, statistics, artificial intelligence, natural language processing,

© Tsinghua University Press 2020
X. Liang, *Social Computing with Artificial Intelligence*,
https://doi.org/10.1007/978-981-15-7760-4_1

machine learning, deep learning techniques, and so on. New analytical tools and technologies have enabled us to make better decisions. However, big data has also brought privacy issues subsequently. For example, in April 2018, Facebook CEO Zuckerberg went to a congressional hearing to receive information from senators about the suspected leak of 87 million user-sensitive information from Facebook to Cambridge Analytics to assist Trump in the election.

In any case, it is worth mentioning that in the field of social computing, researchers utilize massive network data to analyze the opinions of online user groups, excavating huge potential value and providing important help for decision-makers to make more effective decision analysis. This is quite difficult to achieve in traditional research. For example, Wang selected three well-known hot pot chain enterprises as the research object. According to the principle of MapReduce, he collected and processed the sample data, and adopted online analytical processing (OLAP) technology to achieve the visualization of customer network satisfaction in large data environment from three indicators: time, region, and satisfaction. Wang put forward that in the context of big data, banks can analyze and mine potential customers for existing credit card customers according to credit card transaction data, and upgrade credit cards at the same time. Through accurate database analysis, cross-selling can be carried out, new customers can be acquired, and lost users can be recovered. Mallinger and Stefl [2] and Lewis et al. [3] combined traditional methods with big data methods to study how people use data in decision-making, analyze user-generated content in the era of big data. They cannot only play the role of computing methods, cope with large data sets, but also ensure accuracy. Ceron et al. [4] obtained the emotional orientation data of netizens in the 2012 French elections through Twitter and predicted the results of the elections, which proved the better predictive ability of social media. Through collecting the public opinion information of "11.16 school bus accident" in Sina, Tencent and People's Network, Kangwei constructed the network topology map of the incident, and then put forward the guiding strategy of the network public opinion. The Public Opinion Research Laboratory of Shanghai Jiaotong University innovated the research paradigm of public opinion, combined big data mining with social investigation, and constructed a "comprehensive public opinion research framework." On the one hand, it utilized big data mining to find out the relationship between various factors in public opinion events and predicted the development trend of public opinion events. On the other hand, it explored individual's cognition and evaluation of social hotspot issues through social investigation and transforms public opinion research from simplification, unilateralization, and staticization to panoramic, three-dimensional and dynamic.

In conclusion, we believe that confronted with massive online social media data, in order to fully tap the potential value of user groups, social computing future research areas will have the following trend characteristics: (1) Social computing research task needs to shift from the "causal relationship" of the data to the "relevant relationship." In the social computing research activities, we no longer pay too much attention to "causality," but put more emphasis on "correlation" [5].The most typical example is Google Flu Trends, which accurately estimates flu epidemics based on billions of search results around the world [6]. (2) Social computing research objects need to shift from single structured text data to multi-source, unstructured hetero-geneous data. Traditional structured or semi-structured data has gradually turned to unstructured data composed of real-time text, audio, video, etc. Most of them are PB-level multi-source heterogeneous data streams, which are far beyond the scope of literature analysis, comparison, induction, and deduction, and are the contents of "not researching" or "difficult researching" in the past. (3) Social computing research methods need to shift from time-delay, static traditional analysis methods to real-time, dynamic, and interactive big data analysis methods. The traditional "micro-processing" method for small data is obviously incapable of facing the multi-type, multimedia, cross-time, cross-geographic, cross-language big data, and the user's individual needs. In the big data environment, the data analysis process emphasizes real-time, situational, and coordination. The platform for analysis requires higher scalability, fault tolerance, and support for heterogeneous data sources. (4) Social computing research process needs to shift from a "top-down" model to a "bottom-up" model. Traditional empirical research emphasizes the establishment of hypotheses, the collection of data, and the applicability of falsification theory under the premise of theory. It is a top-down analysis model. Random data sampling, data acquisi-tion, verification hypothesis, questions that are not asked in the survey will not be answered, and the research on the target problem of graduate students will not be involved. In the big data environment, users are more urgently required to automati-cally identify valuable rules or patterns from massive data. It is a bottom-up mining model. (5) Social computing research applications need to be transferred from infor-mation decision services to knowledge decision services. Application services in a big data environment require addressing the problem of sparse value in big data, providing the most direct, reliable, real-time, and intelligent visualization solutions for the specific needs of business managers or government decision-makers, from traditional information services to knowledge service. Based on the above analysis, it is known that the innovation of social computing research method in the big data environment is the inevitable requirement for the development of network technology and network resources to a certain scale and degree. It has become an important topic of concern in the academic community and the industry, so it has a strong research significance and research value.

Therefore, this book came into being under this research trend. This book is aimed at social media network as the research object, taking online group analysis and value mining as the research goal, using machine learning-related algorithms to systematically sort out the data processing, method model and practical application of social computing research under big data environment. It should be pointed out that the purpose of this book is to provide a more systematic "data-model-application" theoretical research perspective for researchers in the field of social computing, so only the key areas and achievements of research are mentioned. We hope that readers can be inspired, and carry out further and in-depth research in a certain field under the guidance of a complete theoretical framework.

1.2   Mainstreams of Research Field

At present, the main research fields and methods of social computing focus on social data perception and knowledge discovery, individual and group community modeling, information dissemination and network dynamic evolution, decision support and application and other fields.

1.2.1   Social Data Perception and Knowledge Discovery

Social data acquisition and rule knowledge mining include social learning, social media analysis and information mining, emotion and opinion mining, behavior recognition and prediction. The main forms of social data include text, image, audio, video, etc. Its sources include not only network media information (including blogs, forums, news websites), but also private networks, traditional media, and closed source data of application departments. Knowledge discovery based on social data includes the analysis and mining of social individual or group behavior and psychology. Various learning algorithms have been used to predict organizational behavior. Based on behavioral prediction, the planning reasoning method can identify deep information such as the target and intention of the behavior. The psychological analysis of social groups is mainly oriented to text information (including texts transformed into speech recognition). By analyzing a large amount of social media information, it excavates the opinions and emotional tendencies of netizens.

1.2.2   Community Modeling of Individuals and Groups

Community modeling of individuals and groups includes the construction of behavior, cognitive and psychological models of social individuals or groups, and the analysis of behavior characteristics of social groups. It also includes the modeling of community structure, interaction patterns, and social relations among individuals. Many theoretical models of social sciences are related to the social modeling of individuals and groups. For example, social psychology reveals the formation mechanism of social cognition and psychology and its basic law of development. Social dynamics studies the dynamic process and evolution law of human social development. From the perspective of computating, the study of social individuals and groups is mostly based on textual data, and the trend of recent work is to analyze and model the characteristics of multimedia data and group behavior. Social network is the main means of describing the social interaction and interaction between individuals. The identification of social groups mainly finds potential social groups through the link relationship between network nodes.

1.2.3   Information Dissemination and Dynamic Evolution of Network

The research of information dissemination and network dynamic evolution is to analyze the characteristics of crowd interaction and the evolution of social events, including social network structure, information diffusion and impact, complex network and network dynamics, group interaction and collaboration. Computational sociology [7] holds that a large amount of information on the Internet, such as blogs, forums, chats, consumption records, e-mail, etc., are the mapping of human or organizational behavior in the real world in cyberspace. These network data can be used to analyze the behavior patterns of individuals and groups, thus deepening people's understanding of life, organization, and society. The study of computational sociology involves people's interactions, the form of social group networks, and their evolutionary laws. The evolutionary law analysis of social events mainly focuses on the analysis and evaluation of the process and mechanism of the occurrence, development, intensification, maintenance, and attenuation of social events. For example, in analyzing the evolution law of group activities, researchers use social dynamics to analyze the law of human space-time trajectory based on long-term tracking and detection of 100,000 mobile user terminals, and find that people's operation mode follows repeatable mode [8]. In addition, researchers have used a variety of models to analyze the dissemination, diffusion, and influencing factors of information in the network.

1.2.4   Decision Support and Application

The applications of social computing in the fields of social economy and security include providing decision support, emergency warning, policy evaluation and suggestions to managers and society. In recent years, social computing has made great progress and has been widely used. Network social media can often make a more rapid, sensitive, and accurate response than traditional media because it can fully reflect people's value orientation and true will. Open source information plays an important role in decision support and emergency warning. In the field of social and public security, the Intelligence and Informatics Research Team of the Institute of Automation of the Chinese Academy of Sciences cooperated with relevant national business departments to develop a large-scale open source intelligence acquisition and analysis processing system based on the ACP method, real-time monitoring, analysis and early warning of social intelligence. As well as decision support and services, it has been widely used in the actual business and security related fields of relevant departments. Socio-cultural computing has been applied to security and anti-terrorism decision-making early warning [9, 10]. In addition, due to the complexity of social systems, large-scale social computing research needs computing environment and platform support, including cloud computing platform and various modeling, analysis, application, integration tools, and simulation environment.

1.3   Structure of This Book

This book integrates the four main research fields of social computing with artificial intelligence, namely supervised and unsupervised learning models, social data perception and knowledge discovery, individual and group community modeling, information dissemination and network dynamic evolution, decision support and application, into three research dimensions of data, methods, and applications, and forms the whole content of this book.

Furthermore, the book is divided into three parts: the first part, the data (including Chaps. 2 and 3), introduces the social computing data collection and data analysis algorithm, which roughly corresponds to the social data perception in the four fields of social computing. The second part, model (including Chaps. 4, 5, 6, 7 and 8), introduces supervised, unsupervised, and semi-supervised learning models, as well as more state-of-the-art artificial intelligence algorithms, the opinions content mining, community network structure, and social computing. The three aspects of social computing model of dynamic information dissemination roughly correspond to knowledge discovery, community modeling of individuals and groups, information dissemination, and network dynamic evolution in the four fields of social computing. The third part, application (including Chaps. 9, 10, 11 and 12), introduces the application cases of public security and emergency management, social computing application in business decision support, unsupervised oracle handwriting recognition, and social computing application in online crowd behavior and psychology.

Furthermore, for the specific content of each chapter is concerned, this chapter points out the research background, summarizes the research fields and main methods of social computing. Chapter 2 focuses on data-related content, including the source and classification of data in social computing, data acquisition methods and main tools, as well as data acquisition model and system platform. Chapter 3 introduces data processing principles and methods. Chapter 4 introduces the supervised and unsupervised machine learning method for data analysis. One of the major contributions of this book is to systematically summarize, sort out, and classify the current mainstream machine learning methods, and introduce them according to supervised learning, unsupervised learning and reinforcement learning. In addition, in recent years, deep learning has gained more outstanding performance in the field of social computing. Therefore, we list in-depth learning as a separate section to elaborate. Chapter 5 specifically introduces more state-of-the-art artificial intelligence algorithms, such as deep learning, reinforcement learning, brother learning, and epiphany learning. Chapter 6 introduces public opinion mining and analysis based on social network content mining. According to different content types, this chapter is further divided into text information mining analysis and network image mining analysis. Chapter 7 introduces the research of community discovery based on social network structure in social computing. It is further divided into the introduction of social network topology structure and model and the research of public opinion network community discovery. Chapter 8 introduces user role value mining based on network communication mechanism in social computing. This chapter introduces the analysis of the mechanism of network communication, the model of the influence of public opinion on users' behavior, and the research on the identification of users' roles in social networks. Chapter 9 introduces the application of social computing in emergency management decision support. Chapter 10 introduces the application of social computing in early risk warning in business decision support. Chapter 11 illustrates the unsupervised handwriting recognition method based on pic2vec and also discusses whether oracle bone inscriptions were engraved by right hand or left hand. At the end of the book, Chap. 12 presents social computing applications in online crowd behavior and psychology.

References

1. Lazer D, Pentland A, Adamic L (2009) Computational social science. Science 323(1):721–723
2. Mallinger M, Stefl M (2015) Big data decision making. Graziadio Bus Rev 18(2):25–33
3. Lewis SC, Zamith R, Hermida A (2013) Content analysis in an era of big data: a hybrid approach to computational and manual methods. J Broadcast Electron Media 57(1):34–52
4. Ceron A, Curini L, Iacus SM (2014) Every tweet counts? How sentiment analysis of social media can improve our knowledge of citizens' political preferences with an application to Italy and France. New Media Soc 16(2):340–358
5. Mayer-Schönberger V, Cukier K (2013) Big data: a revolution that will transform how we live, work, and thinking. Houghton Mifflin Harcourt, Boston, pp 50–72
6. Carneiro HA, Mylonakis E (2009) Google trends: a web-based tool for real-time surveillance of disease outbreaks. Clin Infect Dis 49(10):1557–1564
7. Lazer D, Pentland A, Adamic L (2009) Life in the network: the coming age of computational social science. Science 323(5915):721–723
8. González MC, Hidalgo CA, Barabási AL (2009) Understanding individual human mobility patterns. Nature 458(7235)
9. Subrahmanian VS (2007) Computer science. Cultural modeling in real time. Science 317(5844):1509
10. Subrahmanian VS, Albanese M, Martinez MV (2007) CARA: a cultural-reasoning architecture. IEEE Intell Syst 22(2):12–16

Part I
Data

Part I of the book is called *Data*, which includes Chaps. 2 and 3. As we all know, *Data* is a very important research foundation in all social computing studies. Therefore, we intentionally use two chapters to introduce data acquisition and data analysis algorithms in the field of social computing. Specifically, Chap. 2 briefly introduces data sources and classification, as well as data acquisition tools in the field of social computing. In Chap. 3, we address the data processing principles and methods.

Chapter 2
Data Collection

2.1 Data Types and Sources

In the aspect of data acquisition, traditional data acquisition methods, such as questionnaire survey, user interviews, experimental observation, etc., are usually used, which will result in shortcomings such as single data source, small magnitude, scarce type, and delayed information. Even in the Internet age, the method of online questionnaire survey will cause the problem of inaccurate information. In the big data environment, using the characteristics of large data, such as huge volume, variety and fast processing speed, will make up for the above shortcomings, so as to make the data sources of social computing more diverse, more magnitude, richer types, more time-consuming information, and provide better data support for social computing analysis and decision-making in the big data environment. Therefore, due to its extensive research fields and application scenarios, social computing has made its data sources an important feature of mass, diversity, heterogeneity, and timeliness. Generally speaking, social computing involves disparate data including text, images, speech, multimedia social media, and spatial, temporal, and spatiotemporal data.

The main sources of social computing data include communication companies, institutional databases, e-commerce platforms, social networks, search engines, and forums. Its data types include, but are not limited to, telephone communication records, enterprise and government management storage data, e-commerce website transaction data, social network platform and user data, search engine logs, web forum user information and comment data, etc. The data sources and classification of social computing are briefly summarized as shown in Table 2.1.

© Tsinghua University Press 2020
X. Liang, *Social Computing with Artificial Intelligence*,
https://doi.org/10.1007/978-981-15-7760-4_2

Table 2.1 Social computing data sources and classification

Institutional body	Openness	Type
Non-internet institutions	Private data	Transaction data of financial institutions, statistical data of government departments, telephone users' communication records
	Private data	Data published by government business organizations
Internet enterprises	Private data	E-commerce user transaction data, user registration information
	Private data	Social networking user reviews, product reviews, search engine logs, open data source websites

Private data sources are authoritative and confidential to the data, which is usually difficult to obtain. Relevant organizations often use the data processing and analysis department or cooperate with scientific research institutions to explore the value of data. Public data sources, such as Twitter, Facebook, and other social network information, Google search-related data, Amazon product user reviews, etc., often become the focus of various scholars because of their openness, relatively easy access, mass, heterogeneity.

2.2 Data Collection and Tools

2.2.1 Data Acquisition

For the acquisition of public data sources, there are generally three types. The first is to download the processed database directly from the established open database platform for research and application. We bring together several commonly used public databases as shown in Table 2.2. This list of public data sourcesis collected and tidyed from blogs, answers, and user responses. The table below is classified according to domains related to social computing, other extensive awesome sources can be found in this github repository[1]. Researchers can download databases directly from the website for machine learning training and testing (see Table 2.2).

The second is to crawl the information needed for research directly from the webpage. This type of data crawling process is often based on the Cascading Style Sheets (CSS) language to analyze the webpage and then capture the specific data required on the webpage. This type of data grabbing method should often be applied to extract specific data from web pages in the research process, but cannot find ready-made public data sources, and the website does not give the application programming interface website. For example, if a researcher wants to grab the content and comments posted by all users on Reddit from January to June 2018 on SubReddits

---

[1] https://github.com/awesomedata/awesome-public-datasets

called VACCINES (https://www.reddit.com/r/VACCINES/), then he would better crawl the content he needs by using a website crawl.

The third is data capture based on APIs of various websites. Many well-known websites provide dedicated data interfaces for developers and research. The most common websites include Twitter, Microformats, Mailboxes, LinkedIn, Google Buzz, Blogs, Facebook, etc. Researchers can find API documents and rules on relevant official websites. For example, Twitter gives API documents (http://apiwiki. twitter.com/). Through API interfaces, researchers can obtain many needed data. In addition, it should be noted that many researchers have further encapsulated the original API to form a more user-friendly toolkit, many of which are published in github. For example, a minimal wrapper around Twitter's webAPI is available through a package called Twitter (https://github.com/sixohsix/twitter).

2.2.2 Common Data Processing Toolkit

After introducing the sources of data classification and acquisition in the previous section, the next step is to consider whether to use different tools for data processing and analysis to achieve different research purposes. Based on Python programming language, we summarize the data processing toolkits commonly used in social computing data processing, which involve web crawler, text processing, scientific computing, machine learning and in-depth learning, as shown in Table 2.3.

Table 2.2 Commonly used public databases in the field of social computing

Category	Data set	Link	Description
Complex networks	Stanford large network data set collection	http://snap.stanford.edu/data/	SNAP
	Stanford longitudinal network data sources	http://stanford.edu/group/sonia/dataSources/index.html	SoNIA
	Stanford graph base	http://www3.cs.stonybrook.edu/~algorith/implement/graphbase/implement.shtml	–
	Koblenz network collection	http://konect.uni-koblenz.de/	KONECT
	NIST complex networks data collection	https://math.nist.gov/~RPozo/complex_datasets.html	NIST
	Scopus citation database	https://www.elsevier.com/solutions/scopus	SCOPUS
	UCI network data repository	https://networkdata.ics.uci.edu/resources.php	UCINET
	UFL sparse matrix collection	https://sparse.tamu.edu/	SuiteSparse
Data challenges	Kaggle competition data	https://www.kaggle.com/datasets	KAGGLE
	Netflix prize	https://netflixprize.com/leaderboard.html	NETFLIX
	Yelp data set challenge	https://www.yelp.com/dataset	YELP
	Telecom italia big data challenge	https://dandelion.eu/datamine/open-big-data/	Open big data

(continued)

Table 2.2 (continued)

Category	Data set	Link	Description
Machine learning	Travistorrent data set—2017 mining challenge	https://travistorrent.testroots.org/	TravisTorrent
	Keel repository for classification, regression and time series	http://sci2s.ugr.es/keel/datasets.php	Keel
	Machine learning data set repository	http://mldata.org/	MLDATA
	UCI machine learning repository	http://archive.ics.uci.edu/ml/index.php	UCIML
	Million song data set	https://labrosa.ee.columbia.edu/millionsong/	Million song
	Lending club loan data	https://www.lendingclub.com/info/download-data.action	Lending club
Natural language	WordNet	https://wordnet.princeton.edu/	WordNet
	DBpedia	http://wiki.dbpedia.org/Datasets	DBpedia
	Google books ngrams	https://aws.amazon.com/cn/datasets/google-books-ngrams/	Google books
	Gutenberg ebooks list	http://www.gutenberg.org/wiki/Gutenberg:Off line_Catalogs	Gutenberg
	Microsoft machine reading comprehension data set	http://www.msmarco.org/dataset.aspx	MS marco
	Machine translation of European languages	http://statmt.org/wmt11/translation-task.html#download	Machine translation

(continued)

Table 2.2 (continued)

Category	Data set	Link	Description
	Making sense of microposts 2016—named entity recognition	http://micropoccessts2016.seas.upenn.edu/challenge.html	NEEL
	Multidomain sentiment data set (version 2.0)	http://www.cs.jhu.edu/~mdredze/datasets/sentiment/	Multidomain sentiment
	SMS spam collection in English	http://www.dt.fee.unicamp.br/~tiago/smsspamcollection/	SMS spam
	Stanford question answering data set	https://rajpurkar.github.io/SQuAD-explorer/	SQuAD
Online website data sets	Amazon	http://aws.amazon.com/datasets	Amazon
	Reddit data sets	http://reddit.com/r/datasets	Reddit
	Archive.org datasets	https://archive.org/details/datasets	Archive
	Google	http://www.google.com/publicdata/directory	Google
	Data mining and data science	https://www.kdnuggets.com/datasets/index.html	KDNuggets
	Microsoft data science for research	https://www.microsoft.com/en-us/research/academic-program/data-science-microsoft-research/	Microsoft
	Yahoo webscope	https://webscope.sandbox.yahoo.com/	Yahoo
	Washington post list	http://www.washingtonpost.com/wp-srv/metro/data/datapost.html	Washington post
	Wikidata—wikipedia databases	https://www.wikidata.org/wiki/Wikidata:Database_download	WikiData

Table 2.3 Python data processing toolkit related to social computing

Category	Package	Link	Description
Webcrawler toolkit	Scrapy	http://scrapy.org/	Afast high-level screen scraping and web crawling framework for Python
	Beautiful Soup	http://www.crummy.com/software/BeautifulSoup/	Beautifu Soup is not a complete set of crawler tools, it needs to be used with urllib, but a set of HTML/XML data analysis, cleaning and acquisition tools
	Python-Goose	https://github.com/grangier/python-goose	Html content/article extractor, webscrapping lib in Python
Machine learning–data mining	Scikit-learn	http://scikit-learn.org/	It features various classification, regression and clustering algorithms
	Pandas	http://pandas.pydata.org/	Pandas is a software library written for the Python programming language for data manipulation and analysis
	Mlpy	http://mlpy.sourceforge.net/	Mlpy is a Python module for machine learning built on top of NumPy/SciPy and the GNU Scientific Libraries
	MDP	http://mdp-toolkit.sourceforge.net/	Modular toolkit for data processing (MDP) is a Python data processing framework

(continued)

Table 2.3 (continued)

	PyBrain	http://www.pybrain.org/	Its goal is to offer flexible, easy-to-use yet still powerful algorithms for machine learning tasks and a variety of predefined environments to test and compare your algorithms
	NetworkX	https://networkx.github.io/	NetworkX is a graph theory and complex network modeling tool developed in Python language. It has built-in common graph and complex network analysis algorithm, which can easily carry out complex network data analysis, simulation modeling and so on
Deep learning	TensorFlow	https://www.tensorflow.org	TensorFlow is a system that transfers complex data structures to artificial intelligence neural networks for analysis and processing. It can be used in many fields of machine learning and deep learning, such as speech recognition or image recognition
	Caffe	http://caffe.berkeleyvision.org	Convolutional architecture for fast feature embedding is a commonly used deep learning framework, which is widely used in video and image processing

(continued)

Table 2.3 (continued)

Theano	http://www.deeplearning.net/software/theano/	Theano is a Python library that allows you define, optimize, and evaluate mathematical expressions involving multidimensional arrays efficiently
PyTorch	http://pytorch.org/	PyTorch is a deep learning tensor library optimized with GPU and CPU
Pylearn2	http://deeplearning.net/software/pylearn2	Pylearn2 is based on Theano and partly depends on scikit-learning. Pylearn2 will be able to process vector, image, video and other data, and provide in-depth learning models such as MLP, RBM, and SDA

Chapter 3
Data Processing Methodology

3.1 Data Processing Principles

In the existing general big data, structured data only account for about 15%, and the remaining 85% are unstructured data. Under the research of social computing, the proportion of unstructured data will be higher, so it is more necessary to reconsider and design corresponding methods to solve the specific problems encountered in the social computing big data environment. On the one hand, it is necessary to study and discuss through multidisciplinary intersections including mathematics, economics, sociology, computer science, and management science, and to define the individual manifestations, the general characteristics and the basic principles of unstructured and semi-structured data in social computing. On the other hand, each representation of big data only presents the side performance of the data itself, not the whole picture. How to use appropriate data representation and acquisition form for the specific research methods of social computing, and propose a data fusion method for social computing research, will also be an important part of the research. Therefore, after explaining the data classification and source, data acquisition and tools, we will briefly introduce some unique ideas and methods of data acquisition related to social computing in this section.

3.1.1 Behavior Tracking

In the era of big data, people's various behavioral data can be tracked through web logs. The tracking of network user behavior logs is fundamentally different from data acquisition in traditional research. In the past research, researchers mostly used passive questionnaires or surveys to obtain users' data information. This acquisition mode not only makes the amount of data acquired smaller, fewer types, but also makes the reliability of data acquisition relatively low. In the era of big data, we can ensure the timeliness of data to the greatest extent by tracking the historical data of

© Tsinghua University Press 2020
X. Liang, *Social Computing with Artificial Intelligence*,
https://doi.org/10.1007/978-981-15-7760-4_3

user behavior for 7 * 24 h in real time. At the same time, because in many cases, the user cannot feel the existence of data monitoring, so the accuracy of data acquisition can be guaranteed to a certain extent. Therefore, in the era of big data, we will design the corresponding log tracking method for the application of social computing, and match the log collection system to ensure the real-time, reliability, and availability of social computing research data acquisition to the greatest extent.

3.1.1.1   Incremental Acquisition

In traditional research, data acquisition is usually one-time, which leads to a great delay in each research and analysis. Especially with the outbreak of large data, data are constantly generated and updated. Traditional research methods will greatly reduce the effect of their conclusions on the analysis and solution of the latest social problems. Therefore, in view of the social computing research under the big data environment, scholars have proposed the fixed increment crawling method of data. Fixed increment here refers to the acquisition of data at a certain time step, or the acquisition of data when the data produces specific differences. Generally, the acquisition of such data mainly refers to the process of network data collection, obtaining data information from a website through a web crawler or a website public API. This allows unstructured data, semi-structured data to be extracted from webpages and stored in a structured manner as a unified local data file. It supports the collection of pictures, audio, video, and other documents, and the attachments can be automatically associated with the text. For network traffic collection, bandwidth management technologies such as DPI or DFI can be used. Especially at present, online social networks and social media, such as Facebook, Twitter, Sina Weibo and Wechat, are profoundly changing the way people disseminate information and obtain information. Therefore, the design of a targeted method of acquiring and storing data streams in line with the above real-time dynamic new media will provide more abundant and effective data support for the research of social computing.

3.2   Data Processing Methods

Nowadays, with the increasing and diversified social networking platforms, from the user's point of view, everyone often has different corresponding independent accounts for different platforms. From the perspective of network platform, it is often reluctant to share data with other platforms because of the confidentiality and high value of data. However, from the researcher's point of view, synthesizing the diversity data of various platforms is often helpful to depict a user's feature panorama more comprehensively, and then to provide customized services for the user more pertinently. Therefore, in order to achieve interoperability, mutual tolerance, and mutual trust of existing data platforms, matching the behavior data of users in different platforms effectively, and implementing compatibility data acquisition and sharing models and methods for storage platforms of related networks have become a very important research direction in the field of social computing.

Generally speaking, most of the existing data storage platforms use traditional relational databases MySQL and Oracle to store data. However, on the basis of existing research, researchers will use more unstructured databases such as Redis and MongoDB to collect and store data. This method usually deploys a large number of databases at the acquisition end, so it is necessary to study the feasible database architecture model for social computing applications, so as to achieve effective management of these databases. In view of the above requirements of cross-platform heterogeneous databases, it is very important to propose a cross-platform large data processing model. However, at present, the cross-platform large data processing model is still in the exploratory stage based on the data set constraints and data matching difficulties. Cross-platform matching and distributed feature learning are the basis of multi-source and heterogeneous massive data analysis. Key node identification and critical path identification are the key contents of analysis. Scholars' research should break through these aspects.

Part II
AI Models and Algorithms

Part II of the book is called *AI Models and Algorithms*, which includes Chaps. 4–8. In Chaps. 4 and 5, we summarize and sort out the frontier machine learning algorithm, and help readers choose more accurate and effective machine learning algorithm tools when they do research on social computing, including supervised learning, unsupervised learning, semi-supervised learning, deep learning, reinforcement learning, brother learning and epiphany learning. Based on the large amount of heterogeneous data obtained in social media networks, we will further introduce in Part II how to build effective models to solve different research problems in the field of social computing for different research objects and objectives. The user content, topology and user interaction behavior of online social networks constitute three elements of social network. Therefore, the three chapters of Part II will describe the models involved in the analysis of these three elements separately. Specifically, Chap. 6 describes the data mining and knowledge discovery model for online crowd content; Chap. 7 introduces the theoretical basis and model of online crowd community discovery; Chap. 8 introduces the mechanism of information dissemination in social media networks and the model of pattern recognition and influence analysis for online users. The contents of these three chapters will provide effective research tools for readers to study social computing related topics.

Chapter 4
Supervised and Unsupervised Learning Models

In the field of social computing, researchers continue to create more effective algorithms and models, in which machine learning-type algorithm has shown better results, so it has been widely used by researchers. The main contribution of this section is to classify and sort out the advanced machine learning algorithms systematically and comprehensively, introduce the principles, formulas, advantages and disadvantages of each algorithm, and elaborate how some algorithms are applied in the application based on Python language.

In general, machine learning models could be divided into supervised, semi-supervised, unsupervised, and reinforcement learning models. In this chapter, we add a separate section about deep learning only because deep learning algorithms involve both supervised and unsupervised algorithms and they hold a very essential position in current social computing research. Next, we will make a brief review about the principles and applications of these kinds of models respectively.

4.1 Supervised Learning Models

There are still many machine learning models for supervised learning. Researchers can choose the most suitable model to solve practical problems according to the characteristics, advantages, and disadvantages of different models. Generally speaking, supervised machine learning models are mainly used to deal with classification or regression problems. Therefore, in this chapter, we elaborate the principle, classification, regression, advantages, and disadvantages of each model. The application of this chapter is based on Scikit-Learn Python Project published by Pedregosa et al. [1].

© Tsinghua University Press 2020
X. Liang, *Social Computing with Artificial Intelligence*,
https://doi.org/10.1007/978-981-15-7760-4_4

4.1.1   Generalized Linear Algorithms

The model form of the generalized linear algorithms is basically the same, the biggest difference is that the dependent variables are different. If it is continuous, it is linear regression; If it is a binomial distribution, it is a logistic regression;If it is a Poisson distribution, it is a Poisson regression; If it is a negative second-line distribution, it is a negative two-term regression.

According to the classification of use, it can be divided into two categories: regression and classification. The dependent variable is a continuous value, and most of the regression models used for prediction are based on ordinary least squares and ridge regression, lasso, elastic net, etc., which are derived from regularization; the dependent variable is a binomial distribution, which is generally used to classify the model, mainly including logistic regression and softmax.

4.1.1.1   Regression Algorithms

(1)   Ordinary least squares regression (OLSR)

In linear regression model, the target value y is a linear combination of input variables. Therefore, if $\hat{y}$ is the predicted value, the relationship between the input variables and the output y can be showed as below,

$$\hat{y}(w, x) = w_0 + w_1 x_1 + \cdots + w_p x_p,$$

where $w = \left(w_1, \ldots, w_p\right)^{\mathrm{T}}$ is a coefficient, $x = \left(x_1, \ldots, x_p\right)^{\mathrm{T}}$ is the input, and w_0 is the intercept.

Ordinary least squares fits = a linear model with coefficients $w = \left(w_1, \ldots, w_p\right)^{\mathrm{T}}$ to minimize the sum of squared residuals between the observed values in the data set and the responses predicted by linear approximation. Its mathematical formula can be expressed as,

$$\min_{w} ||w^{\mathrm{T}} x - y||_2^2.$$

However, the coefficient estimation of the ordinary least squares method depends on the independence of the model term. When the correlation term and the column of design matrix X have approximate linear correlation, the design matrix becomes nearly singular. As a result, the least square estimation is highly sensitive to the random errors in the observation response and produces large variance. For example, when data are collected without experimental design, this multi-collinearity can be enhanced.

In the probability interpretation of model, the linear model can be explained by its probabilistic meaning. This decomposition will help us better understand the Bayesian regression and Gauss process models described later. We define the true value t_n as input x_n and add a noise generation to the model $y(x, w)$. Its mathematical expression is as follows,

$$t_n = y(x_n, w) + \int \ominus$$

In general, we can define noise $\in$ as Gauss distribution $N(0, \sigma^2)$. Then we can get that t is a Gauss distribution with $y(x, w)$ as its mean,

$$P(t|x, w, \sigma^2) = N(t|y(x, w), \sigma^2), \pi$$

$$b(t|x, w, \sigma^2) = \prod_{n=1}^{N} N(t_n|y(x_n, w), \sigma^2).$$

Then for training data $\{X, t\}$, maximum likelihood estimation can be used to calculate parameters w and σ^2. First, logarithmic likelihood estimation is used,

$$\ln P(t|x, w, \sigma^2) = -\frac{1}{2\sigma^2} \sum_{n=1}^{N} \{y(x_n, w) - t_n\}^2 - \frac{N}{2} \ln 2\pi - N \ln \sigma.$$

The derivation process is as follows,

$$\ln P(t|x, w, \sigma^2) = \ln \prod_{n=1}^{N} N(t_n|y(x_n, w), \sigma^2)$$

$$= \ln \prod_{n=1}^{N} \left\{ \frac{1}{\sqrt{2\pi\sigma^2}} \exp\left(-\frac{\{y(x_n, w) - t_n\}^2}{2\sigma^2}\right) \right\}$$

$$= \ln \left(\frac{1}{\sqrt{2\pi\sigma^2}}\right)^N + \sum_{n=1}^{N} \left(-\frac{\{y(x_n, w) - t_n\}^2}{2\sigma^2}\right)$$

$$= -\frac{1}{2\sigma^2} \sum_{n=1}^{N} \{y(x_n, w) - t_n\}^2 - \frac{N}{2} \ln 2\pi - N \ln \sigma.$$

Next, we estimate the parameter w first, then we can neglect all the items that are not related to w and get the logarithmic likelihood estimation,

$$\ln P(t|x, w, \sigma^2) = \sum_{n=1}^{N} \{y(x_n, w) - t_n\}^2.$$

This similar sum of squares of errors also explains the principle of using sum of squares of errors as a loss function. After estimating w_{ML}, then estimating parameter σ^2, we take $\beta^{-1} = \sigma^2$, then the logarithmic likelihood estimation becomes,

$$\ln P(t|x, w, \beta) = -\frac{\beta}{2} \sum_{n=1}^{N} \{y(x_n, w) - t_n\}^2 - \frac{N}{2} \ln 2\pi - \frac{N}{2} \ln \beta.$$

For its derivative of β, we can get,

$$\frac{\partial \ln P(t|x, w, \beta)}{\partial \beta} = -\frac{1}{2} \sum_{n=1}^{N} \{y(x_n, w) - t_n\}^2 - \frac{N}{2}\beta = 0,$$

$$\sigma_{ML}^2 = \frac{1}{\beta_{ML}} = \frac{1}{N} \sum_{n=1}^{N} \{y(x_n, w_{ML}) - t_n\}^2.$$

Now, the parameters σ_{ML}^2 and w_{ML} have been estimated; we have t's probability distribution model for x,

$$P\left(t|x, w_{ML}, \sigma_{ML}^2\right) = N(t|y(x, w_{ML}), \sigma_{ML}^2).$$

For input x, it is easy to get the corresponding t and its probability.

(2) **Ridge regression**

Ridge regression can solve some problems of ordinary least squares method by imposing penalties on the size of coefficients. The formula for minimizing the sum of squares of penalty residuals of ridge coefficients can be expressed as follows,

$$\min_{w} ||w^T x - y||_2^2 + \alpha ||w||_2^2.$$

Among them, $\alpha \geq 0$ is a complex parameter to control the shrinkage. The larger the value of α, the larger the shrinkage, so the coefficient becomes more robust for the collinearity.

(3) **Least absolute shrinkage and selection operator (LASSO)**

Lasso is a linear model for estimating sparse coefficients. It is useful in some cases because it tends to choose solutions with fewer parameter values, thus effectively reducing the number of variables that a given solution depends on. Lasso and its variants are very important in the field of compressed sensing. In some cases, it can restore the exact combination of nonzero weights.

In mathematical expression, it consists of a linear model trained with L_1 regularity. Its minimization objective function is

$$\min_{w} \frac{1}{2n_{samples}} ||w^T x - y||_2^2 + \alpha ||w||_1.$$

Lasso estimates, therefore, solve the problem of minimizing the least squares penalty by adding $\alpha ||w||_1$, where α is a constant and $\alpha ||w||_1$ is the L_1 regularity of the parameter vector.

(4) Elastic net

Elastic net is a linear regression model trained by L_1 and L_2 regularizations. This combination allows learning sparse models, where weights are mostly zero, like lasso, and ridge's regularization properties are retained. Elastic net is very useful when there are multiple interrelated features. Lasso may randomly select one of the parties involved, while elastic net may pick out both. The objective function of elastic net minimization is

$$\min_{w} \frac{1}{2n_{\text{samples}}} ||w^{\mathsf{T}}x - y||_2^2 + \alpha\rho||w||_1 + \frac{\alpha(1-\rho)}{2}||w||_2^2.$$

One practical advantage of weighing Lasso against ridge is that it allows elastic net to inherit some of ridge's stability in rotation.

(5) Orthogonal matching pursuit (OMP)

Orthogonal matching pursuit achieves approximation of linear model fitting by constraining the number of nonzero coefficients (L_0 pseudo-norm). As a forward feature selection method, OMP can approximate the optimal solution vector with a fixed number of nonzero elements. The formula is expressed as,

$$\arg\min ||y - x\gamma||_2^2 \text{s.t.} ||\gamma||_0 \leq n_{\text{nonzero\_coefs}}.$$

In addition, OMP can focus on specific errors rather than a specific number of nonzero coefficients. This can be expressed as,

$$\arg\min ||\gamma||_0 \text{s.t.} ||y - x\gamma||_2^2 \leq \text{tol}$$

.

OMP is based on greedy algorithm, which contains elements highly related to the current residual at each step. It is similar to the simpler matching pursuit method, but performs better in each iteration. The residuals are recalculated using orthogonal projections in the space of previously selected dictionary elements.

4.1.1.2 Classification Algorithms

(1) Logistic regression

Logistic regression is a classification rather than a nonlinear model. Logistic is also often referred to as logit regression, maximum entropy (MaxEnt) classification or log-linear classifier in literature. In this model, the probability of describing the possible output of a single experiment is modeled by logistic function, and in general, logistic regression is used for bipartition. The prediction function for constructing logistic regression is defined as,

$$w_0 + w_1x_1 + \cdots + w_nx_n = \sum_{i=1}^{n} w_ix_i = w^{\mathrm{T}}x,$$

$$h_w(x) = g(w^{\mathrm{T}}x) = \frac{1}{1 + \exp^{-w^Tx}}.$$

The value of the prediction function $h_w(x)$ represents the probability that the result takes 1, so the probability of classifying the result into category 1 and category 0 for input x is respectively,

$$P(y = 1|x; w) = h_w(x),$$

$$P(y = 0|x; w) = 1 - h_w(x).$$

Based on the maximum likelihood estimation, the loss function $J(\theta)$ is deduced as follows,

$$\cos t(h_w(x), y) = \begin{cases} -\log(h_w(x)), & \text{if } y = 1, \\ -\log(1 - h_w(x)), & \text{if } y = 0, \end{cases}$$

$$J(\theta) = \frac{1}{m} \sum_{i=1}^{m} \cos t(h_w(x_i), y_i)$$

$$= -\frac{1}{m}[\sum_{i=1}^{m} y_i \log h_w(x_i) + (1 - y_i) \log(1 - h_w(x_i))].$$

We usually choose gradient descent method to get the minimum of loss function, and prevent overfitting by L_1 or L_2 regularization method.

(2) **Softmax**

Softmax regression model is a generalization of logistic regression model in multi-classification. It also achieves good results in the classification of MNIST data sets. We know that in logistic regression, the output $y \in \{0,1\}$. In the softmax multi-classification regression model, $y \in \{1, 2, \ldots, k\}$. For a given input x, we estimate the corresponding probability value $P(y = j|x;w)$ for each class j with the prediction function, that is to say, we estimate the probability that x is divided into each category. Therefore, our prediction function will output a k-dimensional vector (the sum of vector elements is 1) to represent the probability values of the K estimates. Specifically, our prediction function $h_w(x)$ is in the following form,

$$h_w(x^{(i)}) = \begin{bmatrix} P(y^{(i)} = 1|x^{(i)}; w) \\ \cdots \\ P(y^{(i)} = k|x^{(i)}; w) \end{bmatrix} = \frac{1}{\sum_{j=1}^{k} \exp^{w_j^{\mathrm{T}}x(i)}} \begin{bmatrix} \exp^{w_1^{\mathrm{T}}x(i)} \\ \cdots \\ \exp^{w_k^{\mathrm{T}}x(i)} \end{bmatrix}.$$

Note that the term $\dfrac{1}{\sum_{j=1}^{k} \exp^{w_j^T x(i)}}$ normalizes the probability distribution so that the sum of all probabilities is 1.

Furthermore, the loss function $J(\theta)$ of softmax is also generalized by the logistic regression,

$$
J(\theta) = -\frac{1}{m}\left[\sum_{i=1}^{m} y^{(i)} \log h_w\left(x^{(i)}\right) + \left(1 - y^{(i)}\right)\log(1 - h_w\left(x^{(i)}\right))\right]
$$

$$
= -\frac{1}{m}\left[\sum_{i=1}^{m}\sum_{j=1}^{k} 1\{y^{(i)} = j\}\log p\left(y^{(i)} = j | x^{(i)}; w\right)\right]
$$

$$
= -\frac{1}{m}\left[\sum_{i=1}^{m}\sum_{j=1}^{k} 1\{y_n^{(i)} = j\}\log \frac{\exp^{w_j^T x(i)}}{\sum_{i=1}^{k} \exp^{w_1^T x(t)}}\right]
$$

For each class i, softmax trains a logistic regression model classifier $h_w^i(x)$ and predicts the probability when $y = i$. When a new input variable x is predicted, each class is predicted separately, and the class with the greatest probability is selected as the classification result, that is,

$$
\max_i h_w^i(x).
$$

4.1.2 Decision Trees

Decision tree (DT) is a non-parametric supervised learning method for classification and regression. Generally, decision tree is a model that predicts the value of target variables by learning simple decision rules inferred from data characteristics. In the classification decision tree with classified values for target variables, leaves represent classification labels and branches represent the connection of features pointing to corresponding class labels. In the regression decision tree, the objective variable is a continuous value. CART (classification and regression tree), ID3 (iterative dichotomiser 3), C4.5 and C5.0 (different versions of a powerful approach) are considered as the most typical decision trees. We will briefly introduce them below.

4.1.2.1 Algorithms Description

(1) CART

CART was proposed in 1984 by L. Breiman, J. Friedman, R. Olshen, and C. Stone. CART is a binary tree using binary segmentation, in which data are cut into left subtree and right subtree, respectively, in each iteration. CART can be used for classification and regression. Compared with ID3, C4.5 and C5.0, it is more widely used. As far as loss function is concerned, CART uses Gini index in classification, as follows,

$$\text{Gini}(D) = 1 - \sum_{i=0}^{n} \left(\frac{Di}{D}\right)^2,$$

$$\text{Gini}(D|A) = \sum_{i=0}^{n} \frac{Di}{D}\text{Gini}(Di).$$

In addition, the loss function of CART regression is mean squared error, which uses the average at the terminal node to minimize L_2 error, and mean absolute error, which uses the median at the terminal node to minimize L_1 error.

Mean squared error is

$$\ell_m = \frac{1}{N_m} \sum_{i \in N_m} y_{i'}$$

$$H(X_m) = \frac{1}{N_m} \sum_{i N_m} (y_i - c_m)^2.$$

Mean absolute error is

$$\overline{y_m} = \frac{1}{N_m} \sum_{i \in N_m} y_{i'}$$

$$H(X_m) = \frac{1}{N_m} \sum_{i N_m} y_i - \overline{y}_m.$$

(2) ID3

ID3 was proposed by Ross Quuinlan in 1986. ID3 can create a multi-path tree, find a classification feature for each node, and generate maximum information gain for the classification target in a greedy way. Data are segmented according to the principle of "maximum information entropy gain." The information entropy gains of data set D and feature A can be calculated as follows.

1. To calculate the empirical entropy $H(D)$ of data set D,

$$-H(D) = \sum_{k=1}^{K} \frac{|C_k|}{|D|} \log_2 \frac{|C_k|}{|D|}$$

2. To calculate the empirical conditional entropy $H(D|A)$ of feature A for data set D,

$$H(D|A) = \sum_{i=1}^{n} \frac{|D_i|}{|D|} H(D_i) = -\sum_{i=1}^{n} \frac{|D_i|}{|D|} \sum_{k=1}^{K} \frac{|D_{ik}|}{|D_i|} \log_2 \frac{|D_{ik}|}{|D_i|}$$

3. To calculate the information gain,

$$g(D, A) = H(D) - H(D|A).$$

Typically, pruning steps are applied after the ID3 tree grows to its maximum size to improve its robustness.

(3) **C4.5**

C4.5 was proposed by Ross Quinlan in 1993 based on ID3. With regard to the disadvantage of ID3, it usually prefers features with more attribute values in terms of information gain. C4.5 uses gain ratio instead of information gain as loss function. In C4.5, partitioning information is introduced to punish features with more attribute values. In addition, C4.5 makes up for ID3's inability to deal with continuous features,

$$\text{Split information}(D, A) = -\sum_{i=1}^{n} \frac{|D_i|}{|D|} \log \frac{|D_i|}{|D|}$$

$$\text{gain ratio}(D, A) = \frac{g(D, A)}{\text{split inf ormation}(D, A)}.$$

(4) **C5.0**

C5.0 is the latest version of Quinlan released under a proprietary license. It uses less memory and builds smaller rule sets with more accuracy than C4.5.

4.1.2.2 Regression and Classification Application

In order to make the reader know more about the application of decision tree algorithm in practical problems, this book refers to the Scikit-Learn project published by Pedregosa et al. [1] and gives specific examples of decision tree algorithm in regression and classification.

Firstly, we use decision tree regressor in scikit-learning to implement the prediction instance of random data sets. In this example, decision tree is used to fit a sine curve and the data set contains noise points. The example sets'max_depth' to 2 and 5, respectively [2].

In addition, we use Iris data set and the decision tree classifier in scikit-learnt to implement a binary classification decision tree, which will help readers better understand the principle of decision tree algorithm. This is an example of classifying two flower varieties in Iris data set, "versicolor'and'virginica" [3].

4.1.2.3 Pros and Cons

(1) Advantages of decision trees

The advantages of decision trees involve:

1. It is simple to understand and explain, and the decision tree model can be imagined.
2. The amount of data that needs to be prepared is small, while other technologies often require large data sets. Virtual variables need to be created to remove incomplete data, but the algorithm cannot accurately predict lost data.
3. The time complexity of the decision tree algorithm (i.e. the prediction data) is the logarithm of the data points used to train the decision tree.
4. Ability to deal with categories of numbers and data (requires corresponding transformations), while data sets analyzed by other algorithms tend to have only one type of variable.
5. Ability to handle multiple output issues.
6. Using a white-box model, if a given situation is observed in a model, the explanation of the condition is easy to interpret Boolean logic. In contrast, in a black-box model (such as artificial neural network), the result may be more difficult to interpret.
7. Statistical tests may be used to validate the model in order to verify the reliability of the model.
8. From the data results, it works well, although its assumptions are somewhat contrary to the real model.

(2) Disadvantages of decision trees

The disadvantages of decision trees involve:

1. Decision tree algorithm learners can create complex trees, but there is no basis for generalization, which is called overfitting. To avoid this problem, the concept of pruning emerged, that is, setting the minimum number of leaf nodes or the maximum depth of the tree.
2. The result of decision tree may be unstable, because a small change in data may lead to the generation of a completely different tree. This problem can be solved by using integrated decision tree.

3. The actual decision tree learning algorithm is based on heuristic algorithm, such as greedy algorithm, to seek the local optimal decision on each node. Such an algorithm cannot guarantee the return of the global optimal decision tree. This can alleviate the need to train multi-tree ensemble learners, where functions and samples are randomly sampled and replaced.
4. There are some concepts that are difficult to understand, because decision trees themselves cannot easily express them, such as XOR checking or reuse.
5. Decision tree learners are likely to create biased trees when some classes are dominant, so it is recommended to train decision trees with balanced data.

4.1.3 Nearest Neighbors

4.1.3.1 Supervised Nearest Neighbor Algorithms

Nearest neighbor learning can provide supervised learning as well as unsupervised learning. Nearest neighbors unsupervised learning is the basis of many other learning methods, especially manifold learning and spectral clustering. Typical nearest neighbor algorithms for unsupervised use include: BallTree, KDTree, and a brute-force algorithm. Here, we mainly introduce the supervised nearest neighbor algorithms.

Generally speaking, supervised nearest neighbor learning can be used for classification and regression, where classification for data with discrete labels and regression for data with continuous labels. Nearest neighbor learning is to find a predetermined number of training samples nearest to the new point and predict labels from them. The number of samples can be a user-defined constant (K-nearest neighbor learning) or vary according to the local density of the point (radius-nearest neighbor learning). Normally, distance can be any metric: the standard Euclidean distance is the most common choice.

The supervised nearest neighbor algorithm is an instance-based or non-generalized learning algorithm: it does not attempt to build a general internal model, but only stores instances of training data. The two most common nearest neighbor algorithms are K-nearest neighbor learning and radius-nearest neighbor learning, of which K-nearest neighbor learning is more common. K-neighbors classifier implements learning based on k nearest neighbors per query point, where k is a user-specified integer value. The radius neighbors classifier implements learning based on the number of neighbors within a fixed radius r of each training point, where r is a user-specified floating-point value.

Despite the simplicity of the algorithm, recent neighbors have succeeded in a large number of classification and regression problems, including handwritten digital or satellite image scenarios. As a non-parametric method, it is usually successful when the decision boundary is very irregular. In the following, we use K-nearest neighbor algorithms classification and regression application examples for readers to better understand and learn.

4.1.3.2 *K*-Nearest Neighbor Classification and Regression Application

In this section, we present an example of using *K*-nearest neighbor classification algorithm to implement three-class classification for Iris data sets, such as "setosa", "versicolor", "virginica." By default, the algorithm uses a uniform weight, that is, weight = "uniform," to indicate that each point in the internal neighborhood plays a role in the classification of classification points. However, in some cases, the nearer the point to be classified, the greater the contribution to the classification is. Therefore, this problem can be solved by weight keywords, that is, when weight = "distance," the weight is proportional to the reciprocal of the distance from the point to be classified. Alternatively, a user-defined distance function can be provided, which will be used to calculate weights. The actual application still refers to the scikit-learning project [3].

In addition to the conventional single output *K*-nearest neighbor regression application, a very classical regression algorithm application is a multi-output regression example. This example uses a multi-output estimator to complete image complementation. The target is to predict the image of the lower half of the face given the upper half of the face. This regression example also compares four algorithms, *K*-nearest neighbor regression, extra trees regression, linear regression, and ridge regression [4].

4.1.3.3 Pros and Cons

Pros:
The theory is mature and the idea is simple. It can be used for both classification and regression.
It can be used for nonlinear classification.
It has no hypothesis about data, high accuracy and is insensitive to outlier.
It is suitable for classifying rare things.
It is especially suitable for multi-classification problems.
Cons:
Lazy algorithm, large amount of calculation;
Sample imbalance problem, such as the large sample size of one category, may lead to the large sample size, which will affect the classification effect when calculating the nearest neighbors of new samples.

To scan all training samples to calculate the distance, the memory overhead is large and the score is slow.
The interpretability is not good enough to give rules like decision tree.

4.1.4 Bayesian Methods

4.1.4.1 Naive Bayes

Naive Bayesian method is a group of supervised learning algorithms, which is a Bayesian theorem based on the "naive" hypothesis. The naive hypothesis is that each pair of features is independent. Specifically, Bayesian theory describes a given set of correlation eigenvectors $\{x_1, x_2, \ldots, x_n\}$ and the corresponding class variable y have the following relations,

$$P(y|x_1, \ldots, x_n) = \frac{P(y)P(x_1, \ldots, x_n|y)}{P(x_1, \ldots, x_n)}.$$

If a simple assumption of independence is given, there is

$$P(x_i|y, x_1, \ldots, x_{i-1}, x_{i+1}, \ldots, x_n) = P(x_i|y).$$

Therefore, the naive Bayesian theorem can be simplified as follows,

$$P(y|x_1, \ldots, x_n) = \frac{P(y) \prod_{i=1}^{n} P(x_i|y)}{P(x_1, \ldots, x_n)}.$$

Since the input $P(x_1, \ldots, x_n)$ is a constant, the classification rules can be expressed as,

$$P(y|x_1, \ldots, x_n) \propto P(y) \prod_{i=1}^{n} P(x_i|y),$$

$$\hat{y} = \arg \max_{y} P(y) \prod_{i=1}^{n} P(x_i|y).$$

We can use the Maximum A Posterior (MAP) estimation to estimate $P(y)$ and $P(x_i|y)$. In supervised learning training, the former represents the relative frequency of category y in the training set. Different naive Bayesian classifiers are mainly based on their assumptions of distribution. Generally, it can be divided into Bernoulli naive Bayes, Gaussiannaive Bayes, Multinomial naive Bayes.

(1) Bernoulli naive Bayes

Bernoulli nave Bayes implements the training and classification algorithm of naive Bayes according to the data of multivariate Bernoulli distribution. That is, data may have multiple features, but each feature is assumed to be binary-valued (Bernoulli, Boolean) variable. Therefore, Bernoulli naive Bayes requires that the input training sample feature variables be binary, and if it is of other types, the algorithm binarizes them.

The decision rules of Bernoulli naive Bayes are as follows,

$$P(x_i|y) = P(i|y)x_i + (1 - P(i|y))(1 - x_i).$$

(2) Multinomial naive Bayes

Multinomial naive Bayes implements naive Bayesian algorithm for polynomial distributed data. The distribution is parameterized for each category y by the vector $\theta_y = (\theta_{y1}, \theta_{y2}, \ldots, \theta_{yn})$, where n represents the number of features and θ_{yi} represents the probability $P(x_i|y)$ that feature i appears in class y samples.

The parameter θ_y refers to the relative frequency counting, which is estimated by a smoothed version of maximum likelihood,

$$\hat{\theta}_{yi} = \frac{N_{yi} + \alpha}{N_y + \alpha n},$$

where $N_{yi} = \Sigma_{x \in T} x_i$ refers to the number of times that feature i appears in the class y sample of training set T, and $N_y = \sum_{i=1}^{|T|} N_{yi}$ refers to the total number of all features of class y. The smoothing prior $\alpha \geq 0$ takes into account the absence of features in the learning samples and prevents zero probability in further calculation. Setting $\alpha = 1$ is called Laplace smoothing, and when $\alpha < 1$ is called Lidstone smoothing.

(3) Gaussian naiveBayes

In the Gaussian naive Bayes algorithm, the likelihood of a hypothesis is subject to the Gaussian distribution,

$$P(x_i|y) = \frac{1}{\sqrt{2\pi\sigma_y^2}} \exp\left(-\frac{(x_i - \mu_y)^2}{2\sigma_y^2}\right).$$

The parameters σ_y and μ_y are obtained by maximum likelihood estimation.

Overall, although the assumptions of naive Bayesian algorithm are too simple, the classifier works well in many real-world situations, which is well known for document classification and spam filtering. They need a small amount of training data to estimate the necessary parameters. Compared with more complex methods, naive Bayesian learners and classifiers can be very fast. Decupling of categorical conditional feature distribution means that each distribution can be independently estimated as a one-dimensional distribution. This, in turn, helps alleviate problems caused by the curse of dimensions. On the other hand, although naive Bayes is considered to be a good classifier, it is not a good estimator, so its probability output value will not be overly valued.

4.1.5 Bayesian Regression

(1) Bayesian priori interpretation of regular terms

In the previous section, we give a probability explanation of linear regression,

$$P(t|x, w, \sigma^2) = N(t|y(x, w), \sigma^2).$$

Let $\beta^{-1} = \sigma^2$ and take a logarithm of the above formula and then get,

$$\ln p(t|x, w, \beta) = -\frac{\beta}{2} \sum_{n=1}^{N} \{y(x_n, w) - t_n\}^2 - \frac{N}{2} \ln 2\pi - \frac{N}{2} \ln \beta,$$

where the true value t is the input x is added to the model $y(x, w)$ by a noise, the noise ϵ is the Gaussian distribution $N(0, \sigma^2)$, and t is the mean of Gaussian distribution $y(x, w)$. We further introduce the Bayesian rule, which can be assumed that the parameter w has a Gaussian prior distribution and the variance is α^{-1},

$$P(w|\alpha) = N(w|0, \alpha^{-1}I) = \left(\frac{\alpha}{2\pi}\right)^{(M+1)/2} \exp\left\{-\frac{\alpha}{2}w^T w\right\},$$

where $M + 1$ represents the complexity of the model, i.e., the number of polynomial regressions. Then, according to the Bayesian rules, there is

$$P(w|x, t, \alpha, \beta) = P(t|x, w, \beta)P(w, \alpha).$$

This is called the Maximum Posterior of MAP. After removing the irrelevant terms, the logarithmic likelihood can be obtained as follows,

$$\ln P(w|x, t, \alpha, \beta) = \ln P(t|x, w, \beta) + \ln P(w, \alpha)$$

$$= -\frac{\beta}{2} \sum_{n=1}^{N} \{y(x_n, w) - t_n\}^2 + \frac{\alpha}{2}w^T w.$$

It can be seen that the prior probability corresponds to the regular term whose regularization parameter is $\lambda = \alpha/\beta$. It can be guessed that the complex model has a smaller prior probability, while the relatively simple model has a larger prior probability. Then the regularization used by L_2 in ridge regression is equivalent to finding a maximum delay solution under Gaussian priori with a parameter w and an accuracy of λ^{-1}. And it is not a manual setting of λ, it is possible to take it as a random variable to estimate from the data.

(2) **Bayesian linear regression**

The model is a linear function of the parameter vector $w = (w_0, w_1, w_2, \ldots, w_M)^{\mathrm{T}}$ and also represents a linear function of the input variable x, but this greatly limits the applicability of the model. Therefore, we use the basis function to extend the above linear model, that is, the linear regression model is a linear combination of a set of nonlinear basis functions of input variable x, which is expressed mathematically as follows,

$$y(x, w) = \sum_{j=0}^{M} w_j \varphi_j(x) = w^{\mathrm{T}} \varphi(x)$$

$$w = (w_0, w_1, w_2, \ldots, w_M)^{\mathrm{T}}$$

$$\varphi = (\varphi_0, \varphi_1, \varphi_2, \ldots, \varphi_M)^{\mathrm{T}}.$$

Among them, $\varphi_j(x)$ denotes the basis function, and the total number of basis functions is M, let $\varphi_0(x) = 1$. Let t be the target output corresponding to x, and β^{-1} be the variance of the Gaussian distribution of the sample set. w satisfies the Gaussian distribution, and α^{-1} is the variance of the parameter w Gaussian distribution. From the above analysis, we can get the probability representation of the linear model as follows. Generally, we call $P(w)$ conjugate prior,

$$P(t|x, w, \beta) = N\big(t|y(x, w), \beta^{-1}I\big),$$

$$P(w|\alpha) = N\big(w|0, \alpha^{-1}I\big) = P(w).$$

Then the logarithmic posteriori probability function of the linear model is as follows, which is also called Bayesian ridge regression. It has all the characteristics of ridge regression,

$$\ln P(\theta|D) = \ln P(w|T)$$

$$= -\frac{\beta}{2} \sum_{n=1}^{N} \{y(x_n, w) - t_n\}^2 + \frac{\alpha}{2} w^{\mathrm{T}} w + \text{const.}$$

Here, it is the target value vector of the data sample, D is the sample set, and $T = \{t_1, t_2, \ldots, t_n\}$, const is a quantity independent of parameter w.

(3) **Bayesian linear regression learning process**

Let there be n samples in sample set D, which is denoted as $D^n = \{x_1, x_2, \ldots, x_n\}$. In the case of $n > 1$,

$$P\big(D^n|\theta\big) = P(x_n|\theta) P\big(D^{n-1}|\theta\big),$$

$$P(\theta|D^n) = \frac{P(D^n|\theta)P(\theta)}{P(D^n)} = \frac{P(x_n|\theta)P(D^{n-1}|\theta)P(\theta)}{\int P(x_n|\theta)P(D^{n-1}|\theta)P(\theta)d\theta}.$$

The Bayesian learning process can be described as follows: on the posterior probability $P(\theta|D^{n-1})$ of the previous training set D^{n-1}, multiplied by the likelihood estimate of the new test sample point x_n, the posterior probability $P(\theta|N^d)$ of the new set D^n can be obtained, which is equivalent to the prior probability distribution of $P(\theta|D^{n-1})$,

$$P(\theta|D^n) \propto P(x_n|\theta)P(\theta|D^{n-1}).$$

We describe the learning process of Bayesian linear regression in detail. The model is

$$y(x, w) = w_0 + w_1 x.$$

The first line:
The first line is the initial state, so there is no likelihood estimation about the sample. So, there is no graph on the left side of the first line. There is only prior information about w, i.e., $P(\theta|D^0) = N(w|0, \alpha^{-1}I)$, so the graph is composed of circles centered on $(0, 0)$. Random lines drawn by random selection of points (w_0, w_1) are shown in the rightmost figure of the first row.

The second line:
When the first sample point x_1 enters, the likelihood estimate $P(x_1|\theta)$ of x_1 can be obtained from x_1. It is shown in the first column of the second row that the result of the likelihood estimate is actually a duality of the formula, i.e.,

$$w_1 = \frac{1}{x_1}y - \frac{1}{x_1}w_0.$$

It can be estimated from the right-most data space graph of the second row that the coordinates of the first sample point are approximately $(0.9, 0.1)$. Therefore, in the first graph of the second row, the central line equation of likelihood estimation is

$$w_1 = \frac{1}{9} - \frac{10}{9}w_0.$$

Because the prior distribution of the second line is the posterior distribution of the first line, that is, the picture in the middle of the first line. Then the posteriori distribution of the second line is obtained by multiplying the left side of the second line with the middle one of the first line. The image on the right of the second line is a line drawn by randomly extracting some points (w_0, w_1) from the image in the middle of the second line and taking (w_0, w_1) as a parameter.

4.1.6 Gaussian Processes

Gauss process is a machine learning method based on statistical learning theory and Bayesian theory. It is suitable for dealing with complex regression problems such as high dimension, small sample size, and nonlinearity, and has strong generalization ability. Compared with neural network and support vector machine (SVM), GP has the advantages of easy realization, adaptive acquisition of hyperparameters, flexible non-parametric inference, and probability significance of output.

The Gaussian process extracts a finite number of indices (such as $n, t_1, \ldots, t_n$) from a random variable cluster and the joint distribution of the vectors $(Xt_1, \ldots, Tt_n)$ is a multidimensional Gaussian distribution. In a Gaussian process, each point in the input space is associated with a random variable subject to a Gaussian distribution, and the joint probability of a combination of any finite number of these random variables is also subject to a Gaussian distribution. When the indication vector t is two-dimensional or multidimensional, the Gaussian process becomes a Gaussian random field GRF. The depiction of the Gaussian process, like the Gaussian distribution, is also characterized by mean and variance. Generally, in the method of applying the Gaussian process, it is assumed that the mean m is zero, and the covariance function K is determined according to the specific application.

As a machine learning algorithm, the Gaussian process uses inert learning and point-to-point similarity metrics (kernel functions) to predict the values of invisible points in the training data [5]. The prediction is not just an estimate of the point, but also contains uncertainty information—it is a one-dimensional Gaussian distribution (this is the marginal distribution of the point).

4.1.6.1 Gaussian Processes Algorithms

Gauss process regression modeling stage process can be described as: given training data $x_1, x_2, \ldots, x_n$, the corresponding function values are $y_1, y_2, \ldots, y_n$. Assuming that the observed t is modeled as an objective function $y(x)$ with Gauss noise $t = y(x) + N(0, \beta^{-1})$, the joint probability distribution of the target variable t is as follows,

$$P(t|y) = N(t|y,^{-1} I_N).$$

According to the definition of Gauss process, the edge distribution of $P(y)$ is as follows, $P(y) = N(y|0, k)$, where the covariance is defined by K matrix. The edge distribution of $P(t)$ is obtained as follows, $P(t) = \int P(t|y)P(y)dy = N(t|0, C)$. The elements of covariance matrix C are

$$C(x_n, x_m) = k(x_n, x_m) + \beta^{-1}\delta_{nm}.$$

The prediction stage of Gauss process regression is described as: the task of prediction is to get the distribution of the prediction variable t^* given a new input x^*.

First, we get the joint distribution of $t_{N+1} = \{t_1, t_2, \ldots, t_n, t^*\}$. The covariance matrix is $C_{N+1} = \begin{pmatrix} C_N & k \\ k^T & c \end{pmatrix}$.

According to the conditional distribution formula of multidimensional Gauss distribution, $P(t^* | t)$ can be obtained,

$$P(t^* | t) = N(k^T C_N^{-1} t, c - k^T C_N^{-1} k).$$

So, we use the mean of the distribution as the estimation value.

In the above process, the covariance matrix is usually calculated by the kernel function, which is the key component to determine the prior and posterior shape of GP. By defining the "similarity" of two data points and assuming that similar data points should have similar target values, they encode the hypothesis of the learning function. The kernel functions commonly used in Gauss processes include: basic kernels, radial-basis function (RBF) kernel, matern kernel, rational quadratic kernel, exp-sine-squared kernel, dot product kernel.

(1) **Basic kernels**

Constant kernel can be used as part of product kernel to expand the size of other factors (kernel) or as part of sum kernel, which changes the average value of the Gauss process. It can be defined as,

$$k(x_i, x_j) = \text{const}, \quad \forall x_1, x_2.$$

A major example of white kernel's use is that it explains the noise part of the signal as part of the sum-kernel. It can be defined as,

$$k(x_i, x_j) = \text{noise level if } x_i = x_j \quad \text{else } 0.$$

1. RBF kernel

The RBF kernel is also known as the "squared index" kernel. It is parameterized by the length parameter $l > 0$, which can be a scalar (the isotropic variant of the kernel), or a vector with the input x dimension (the anisotropic variant of the kernel), which can be expressed as,

$$k(x_i, x_j) = \exp\left(-\frac{1}{2} d\left(\frac{x_i}{l}, \frac{x_j}{l}\right)^2\right).$$

This verification is infinitely divisible, which indicates that the Gaussian process using the RBF kernel as the covariance has a mean square derivative of all orders and is therefore very smooth.

2. Matern kernel

Matern kernel is a generalization of the RBF kernel, which uses an additional parameter v to control the smoothness of the resulting function. Like the RBF, it is parameterized by the length parameter $I > 0$, which can be a scalar (the isotropic variant of kernel) or a vector with the input x dimension (the anisotropic variant of the kernel). Matern kernel can be expressed as,

$$k(x_i, x_j) = \sigma^2 \frac{1}{\Gamma(v)2^{v-1}} \left(\gamma \sqrt{2v} d\left(\frac{x_i}{l}, \frac{x_j}{l}\right) \right)^v K v \left(\gamma \sqrt{2v} d\left(\frac{x_i}{l}, \frac{x_j}{l}\right) \right).$$

When $v \to \infty$, matern kernel converges to the RBF kernel. When $v = 1/2$, the Matern kernel is equal to the absolute exponential kernel,

$$k(x_i, x_j) = \sigma^2 \exp\left(-\gamma d\left(\frac{x_i}{l}, \frac{x_j}{l}\right) \right) v = 1/2.$$

Especially when $v = 2/3$,

$$k(x_i, x_j) = \sigma^2 \left(1 + \gamma \sqrt{3} d\left(\frac{x_i}{l}, \frac{x_j}{l}\right) \right) \exp\left(-\gamma \sqrt{3} d\left(\frac{x_i}{l}, \frac{x_j}{l}\right) \right) v = 2/3.$$

When $v = 5/2$,

$$k(x_i, x_j)$$
$$= \sigma^2 \left(1 + \gamma \sqrt{5} d\left(\frac{x_i}{l}, \frac{x_j}{l}\right) + \frac{5}{3} \gamma^2 d\left(\frac{x_i}{l}, \frac{x_j}{l}\right)^2 \right) \exp\left(-\gamma \sqrt{5} d\left(\frac{x_i}{l}, \frac{x_j}{l}\right) \right) v = 5/2.$$

These are the most mainstream choices in the learning process. These functions are not infinitely divisible (as assumed by the RBF kernel), but at least once ($v = 2/3$) or twice differentiable ($v = 5/2$).

(2) **Rational quadratic kernel**

The rational quadratic kernel can be thought of as a proportional mixture (infinite kernel) of RBF kernels with different feature length scales. It is parameterized by the length parameter $I > 0$ and the proportional mixing parameter $\alpha > 0$. Only l is a scalar isotropic variable at this time. The kernel is given by

$$k(x_i, x_j) = \left(1 + \frac{d(x_i, x_j)^2}{2\alpha} \right)^{-\alpha}.$$

1. Exp-sine-squared Kernel

Exp-sine-squared kernel allows modeling of periodic functions. It passes the length parameter $I > 0$ and the periodic parameter $p > 0$. Currently only l is a scalar isotropic variant. The kernel is given by

$$k(x_i, x_j) = \exp\left(-2\left(\sin\left(\frac{\pi}{p}d(x_i, x_j)\right)\right)^2\right).$$

2. Dot product kernel

The dot product kernel can be obtained by placing $N(0, 1)$ priori on the coefficient $x_d(d = 1, 2, \ldots, D)$ in the linear regression and placing $N(0, \sigma_0^2)$ prior on the bias. The dot product kernel rotates unchanged on the origin coordinates. It is parameterized by σ_0^2. When $\sigma_0^2 = 0$, the kernel is called the homogeneous linear kernel; otherwise, it is called inhomogeneous. It can be expressed as,

$$k(x_i, x_j) = \sigma_0^2 + x_i \cdot x_j.$$

4.1.6.2 Gaussian Processes Regression (GPR)

In order to achieve the regression goal of the Gauss process, the GP priori needs to be formulated first. The average value of a priori is set to a constant of 0 or the average value of the training set. The covariance of a priori is specified by passing a kernel object. In the fitting process of the Gaussian process regressor, the hyperparameters of the kernel are optimized by maximizing log-marginal-likelihood (LML) by passing optimizer. Because LML may have multiple local optima, the optimization program can be repeated by specifying parameters. The first run always starts with the initial superparametric values, and then the superparametric values of the rounds are selected randomly from the allowable range. If the initial hyperparameter remains fixed, there is no passing optimizer.

The noise in the target can be specified by transferring the parameter α to the global scalar or to each data point. It is worth noting that the appropriate noise level will also be helpful in dealing with the numerical problems in the fitting process, because it effectively solves the Tikhonov regularization, that is, by adding it to the diagonal line of the kernel matrix. Another way to explicitly specify the noise level is to add a white kernel, which can estimate the global noise level in the data.

For example, the following figure shows a Gaussian process [6] using a sum-kernel containing RBF kernel and white kernel to estimate the noise level of the data. An illustration of the LML landscape indicates the presence of local maxima for two LMLs. The first corresponds to a model with a high noise level and a large length scale, which explains all the changes in the noise data. The second has a smaller noise level and a shorter length scale, which explains most of the changes in the noiseless function relationship. The second mode has a higher probability; however, based on the initial value of the hyperparameter, gradient-based optimization may also converge to a high noise solution. So, it is very important to repeat multiple optimizations for different initializations.

4.1.6.3   Gaussian Processes Classification (GPC)

The classification of Gauss process is actually a probabilistic classification problem, in which the test prediction takes the form of class possibility. Gaussian classification places the Gaussian priori on the potential function f and then obtains the probability Classification by flattening the priori through a link function. Latent functions are called damage functions, and their values are not observed, nor are they related to themselves. The purpose is to facilitate the expression of the model and to be removed (or integrated) in the prediction process. Gauss regression classifier realizes logical link function. Its integral cannot be calculated by analysis, but it is easy to approximate calculation in binary case.

Compared with regression settings, even if the prior conforms to the Gauss distribution, the posterior of potential function f is not Gauss, because the Gauss possibility is not applicable in discrete class labels. Instead, the non-Gauss likelihood corresponding to the logical link function is used. GPC estimates non-Gauss posteriori using Gauss estimation based on Laplacian estimation. GPC can be used for one-vs-one binary classification and one-vs-rest multi-class classification.

4.1.6.4   Pros and Cons

Advantages of the Gauss process:

1. Predictions are interpolations of observations (at least for ordinary kernels).
2. The predicted value is probability (Gauss), so we can calculate the empirical confidence interval, and then according to this information, we can refit (online fitting, adaptive fitting) the prediction in a region of interest.
3. Multifunction: Different kernels can be specified. Common kernels are provided, but specific kernels can also be specified.

Disadvantages of the Gauss process:

1. Not sparse. That is to say, they use complete sample/feature information for prediction.
2. Loss of effectiveness in high-dimensional space—when the number of features exceeds dozens.

4.1.7   Support Vector Machines

4.1.7.1   SVM Algorithm

SVM means a classifier based on support vector operations, is a supervised model that can be used for classification and regression. In the past ten years, because of its good performance, it has become one of the most popular algorithms in the field of machine learning. In order to elaborate SVM in more detail, we describe it as three kinds of models in the order from simplicity to complexity, namely linear SVM for linearly separable case and nonlinear SVM. Its complexity increases in turn. Simple models are the basis of complex models and the special case of complex models.

Simply put, when the training data set is linearly separable, we use the linear branching support vector machine to find a separation surface in the feature space and divide the instances into different classes by maximizing the hard margin, but this method is not applicable to the training data set which is linearly inseparable; when the training data set is linearly inseparable, and after removing some specific points, most of the sample points are set. When the sum is linearly separable, we use linear support vector machine to solve a convex quadratic programming problem by maximizing the soft margin. When the classification problem is nonlinear, we use the nonlinear support vector machine to transform the nonlinear problem into a linear problem by using the kernel technique and then get the ideal result. We will introduce these three models in turn in the following sections.

(1) **Linear support vector machine in linearly separable case and hard spacing maximization**

Assume that given a linearly separable training data set in feature space,

$$T = \{(x_1, y_1), (x_2, y_2), \ldots, (x_N, y_N)\},$$

where $x_i \in X = R^m$, $y_i \in Y = \{+1, -1\}$, $i = 1, 2, \ldots, N$, x_i is the ith eigenvector, which is also an example, and y_i is a tag class of x_i. When $y_i = +1$, x_i is called a positive example; when $y_i = -1$, x_i is called a negative example, and (x_i, y_i) is called a sample point. The goal of learning is to find a separate hyperplane in the feature space and divide the instances into different classes. The separation hyperplane corresponds to the equation $w \cdot x + b = 0$, which is determined by the normal vector w and the intercept b, and can be represented by (w, b).

Then the linearly separable support vector machine is defined as follows: Given a linearly separable data set, the isolated hyperplane learned by maximizing or equivalently solving the corresponding convex quadratic programming problem is

$$w^* \cdot x + b^* = 0.$$

The corresponding classification decision function is

$$f(x) = \text{sign}(w^* \cdot x + b^*).$$

It is called linearly separable support vector machine (see Fig. 4.1).

The above figure can help us understand that, in general, when the training data set is linearly separable, there are infinitely separated hyperplanes that can correctly separate the two types of data, but the separated hyperplanes with the largest geometric spacing are unique, so linearly separable support. The vector machine uses the hard-margin maximization principle to find the unique separating hyperplane. The intuitive interpretation of hard margin maximization is that finding the hyperplane with the

Fig. 4.1 Example of a
linearly separable data set
classification

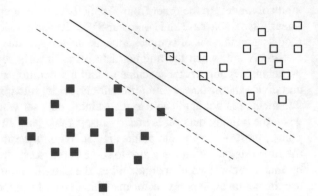

largest margin for the training data set means classifying the training data with sufficient confidence. That is to say, not only the positive and negative instance points
are separated, but also the most difficult instance points (the nearest point to the
hyperplane) are separated with sufficient confidence.

(2) **Maximal margin method**

Therefore, the problem of maximally spaced separating hyperplanes is expressed
mathematically as a constrained optimization problem,

$$\max_{w,b} \gamma,$$

$$\text{s.t. } y_i \left(\frac{w}{||w||} \cdot x_i + \frac{b}{||w||} \right) \geq \gamma, \quad i = 1, 2, \ldots, N.$$

That is to say, we want to maximize the geometric margin γ of hyperplane
(w, b) with respect to training data set, and the constraint condition is that the margin
of hyperplane (w, b) with respect to each training sample point is at least γ. For better
computation, we transform the above formula into an equivalent convex quadratic
programming problem,

$$\min_{u,b} \frac{1}{2}||w||^2,$$

$$\text{s.t. } y_i (w \cdot x_i + b) - 1 \geq 0, \quad i = 1, 2, \ldots, N.$$

If the solution w^* and b^* of the convex quadratic programming problem is obtained,
then the maximum separation hyperplane $w^* \cdot x + b^* = 0$ and the classification
decision function $f(x) = \text{sign}(w^*x + b^*)$, i.e., the linear branching support vector
machine model, can be obtained.

(3) **Dual learning algorithms**

In addition, in order to solve the optimization problem of linearly separable support vector machine, in addition to using the above maximum margin method to obtain the optimal solution, it can also be realized by the dual learning algorithm. The Lagrangian duality is applied to the dual algorithm, and the optimal solution of the primal problem is obtained by solving the dual problem. The advantage of this is that the dual problem is often easier to solve; instead, the kernel function is naturally introduced and then extended to the nonlinear classification problem.

We review the original constrained optimization problem,

$$\min_{x_y, b} \frac{1}{2} \|w\|^2$$

$$\text{s.t. } y_i (w \cdot x_i + b) - 1 \geq 0, \quad i = 1, 2, \ldots, N.$$

First establish the Lagrange function. To this end, for each inequality constraint $y_i (w \cdot x_i + b) - 1 \geq 0, \quad i = 1, 2, \ldots, N$, introducing Lagrange multiplier $\alpha_i \geq 0, i = 1, 2, \ldots, N$. Define the Lagrange function,

$$L(w, b, \alpha) = \frac{1}{2} \|w\|^2 - \sum_{i=1}^{N} \alpha_i \sum_{i=1}^{N} \alpha_i y_i (w \cdot x_i + b) + \sum_{i=1}^{N} \alpha_i.$$

According to Lagrange's duality, the dual problem of the original problem is a very small problem.

Where $\alpha^* = (\alpha_1^*, \alpha_2^*, \ldots, \alpha_N^*)^{\mathrm{T}}$ is a Lagrangian multiplier vector. It can be seen that the Karush-Kuhn-Tucker conditions (KTT) condition is established. We use the KTT condition to solve the problem,

$$\nabla_w L(w^*, b^*, \alpha^*) = w^* - \sum_{i=1}^{N} \alpha_i^* y_i x_i = 0,$$

$$\nabla_b L(w^*, b^*, \alpha^*) = -\sum_{i=1}^{N} \alpha_i^* y_i = 0,$$

$$\alpha_i^* (y_i (w^* \cdot x_i + b^*) - 1) = 0, \quad i = 1, 2, \ldots, N,$$

$$y_i (w^* \cdot x_i + b^*) \geq 0, \quad i = 1, 2, \ldots, N,$$

$$\alpha_i^* \geq 0, \quad i = 1, 2, \ldots, N.$$

Therefore,

$$w^* = \sum_{i=1}^{N} \alpha_i^* y_i x_i,$$

$$b^* = y_i - \sum_{i=1}^{N} \alpha_i^* y_i (x_i \cdot x_j).$$

So, the separation hyperplane can be written as,

$$\sum_{i=1}^{N} \alpha_i^* y_i (x_i \cdot x_j) + b^* = 0.$$

The classification decision function can be written as,

$$f(x) = \text{sign}(\sum_{i=1}^{N} \alpha_i^* y_i (x_i \cdot x_j) + b^*).$$

That is to say, the classification decision function only depends on the inner product of the input x and the input of the training sample. The above formula is called the dual form of the linear branching support vector machine.

We summarize the maximum margin method and dual method as shown in Table 4.1.

(4) **Linear support vector machine and soft margin maximization**

Support vector machine learning method for linearly separable problems is not applicable to linearly separable training data, because inequality constraints in the above methods cannot be established. Linear SVM (LSVM) is suitable for linearly inseparable data. When outlier is removed from the training data set, the set composed of most of the remaining sample points is linearly separable. In order to solve this problem, we can introduce a relaxation variable $\xi_i \geq 0$ for each sample point (x^i, y^i), so that the margin plus relaxation variable is greater than or equal to 1. In this way, the constraint condition becomes,

$$y_i (w \cdot x_i + b) \geq 1 - \xi_i, \quad i = 1, 2, \ldots, N.$$

At the same time, for each relaxation variable ξ_i, a cost ξ_i is paid. The objective function changes from $\frac{1}{2}||w||^2$ to,

$$\frac{1}{2}||w||^2 + C \sum_{i=1}^{N} \xi_i.$$

Table 4.1 Maximum margin method and dual method

Input: Linearly separable training data set $T = \{(x_1, y_1), (x_2, y_2), \ldots, (x_N, y_N)\}$, where $x_i \in X = R^m$, $y_i \in Y = \{+1, -1\}$, $i = 1, 2, \ldots, N$	
Output: separating hyperplanes and separating decision functions	
Maximum margin algorithm process	Dual learning process
Construct and solve constrained optimization problems, $$\min_{x,b} \frac{1}{2}\|w\|^2$$ s.t. $y_i(w \cdot x_i + b) - 1 \geq 0, i = 1, 2, \ldots, N$ The optimal solutions w^*, b^* are obtained. The separating hyperplanes are obtained as follows, $$w^* \cdot x + b^* = 0$$ with the classification decision function, $$f(x) = \text{sign}(w^* \cdot x + b^*).$$	Construct and solve constrained optimization problems, $$\min_\alpha \frac{1}{2}\sum_{i=1}^{N}\sum_{j=1}^{N}\alpha_i\alpha_j y_i y_j (x_i \cdot x_j) - \sum_{i=1}^{N}\alpha_i,$$ s.t. $\sum_{i=1}^{N}\alpha_i y_i = 0,$ $\alpha_i \geq 0, i = 1, 2, \ldots, N$ The optimal solutions $\alpha^* = (\alpha_1^*, \alpha_2^*, \ldots, \alpha_N^*)^T$ are obtained. Calculate $$w^* = \sum_{i=1}^{N}\alpha_i^* y_i x_i.$$ Choose a positive component $\alpha_i^* > 0$ of α^*, calculate $$b^* = y_i - \sum_{i=1}^{N}\alpha_i^* y_i (x_i \cdot x_j).$$ (3) The separating hyperplanes are obtained as follows, $$\sum_{i=1}^{N}\alpha_i^* y_i (x_i \cdot x_j) + b^* = 0$$ with the classification decision function, $$f(x) = \text{sign}\left(\sum_{i=1}^{N}\alpha_i^* y_i (x_i \cdot x_j) + b^*\right).$$

Here, $C > 0$ is called the penalty parameter, which is usually solved by the application problem. The bigger the C value is, the bigger the penalty for misclassification will be. The smaller the C value is, the smaller the penalty for misclassification will be. Then the meaning of the objective function of the minimization of the upper formula is: to make $\frac{1}{2}\|w\|^2$ as small as possible, that is, to maximize the margin, and to minimize the number of misclassified points. C is the coefficients that harmonize the two. This idea is called soft margin maximization.

Then the learning problem of linear inseparable linear support vector machines becomes the following convex quadratic programming problem,

$$\min_{w,b,\xi} \frac{1}{2}\|w\|^2 + C\sum_{i=1}^{N}\xi_i,$$

$$\text{s.t.} y_i (w \cdot x_i + b) \geq 1 - \xi_i, \quad i = 1, 2, \ldots, N,$$
$$\xi_i \geq 0 \quad i = 1, 2, \ldots, N.$$

Then we define linear support vector machine as:

For a given linearly inseparable training data set, by solving the convex quadratic programming problem, i.e., the problem of maximizing the soft margin, the separable hyperplanes are obtained as follows,

$$w^* \cdot x + b^* = 0$$

.

The corresponding classification decision function

$$f(x) = \text{sign}(w^* \cdot x + b^*)$$

is called linear support vector machine. Similarly, we present a dual learning algorithm for linear support vector machines.

Input: Training data set $T = \{(x_1, y_1), (x_2, y_2), \ldots, (x_N, y_N)\}$, where $x_i \in X = R^m$, $y_i \in Y = \{+1, -1\}$, $i = 1, 2, \ldots, N$.

Output: Separating hyperplane and separating decision function.

Process:

1. Choosing penalty parameter $C > 0$ to construct and solve convex quadratic programming problem

$$\min_{\alpha} \frac{1}{2} \sum_{i=1}^{N} \sum_{j=1}^{N} \alpha_i \alpha_j y_i y_j (x_i \cdot x_j) - \sum_{i=1}^{N} \alpha_i,$$

$$\text{s.t.} \sum_{i=1}^{N} \alpha_i y_i = 0,$$

$$C \geq \alpha_i \geq 0, \quad i = 1, 2, \ldots, N$$

.

Obtaining the optimal solution $\alpha^* = (\alpha_1^*, \alpha_2^*, \ldots, \alpha_N^*)^T$.

2. Calculating

$$w^* = \sum_{i=1}^{N} \alpha_i^* y_i x_i.$$

Selecting a component α_i^* of α^* to satisfy $C > \alpha_i^* > 0$, and calculating

$$b^* = y_i - \sum_{i=1}^{N} \alpha_i^* y_i (x_i \cdot x_j).$$

3. Finding the separation hyperplane

$$w^* \cdot x + b^* = 0.$$

Classification decision function is

$$f(x) = \text{sign}(w^* \cdot x + b^*).$$

It can be known from the verification that the interpretation of w is unique, and the solution of b is not unique, so the average value of all the eligible sample points can be taken in the actual calculation.

(5) Nonlinear support vector machine and kernel function

Solving the nonlinear classification problem by linear classification method is divided into two steps: firstly, a transformation is used to map the data of the original space to the new space; then the linear classification learning method is used to learn the classification model from the training data in the new space. Nuclear techniques fall into this category. The basic idea of applying kernel technique to support vector machines is to correspond input space (Euclidean space R^m or discrete set) to a feature space (Hilbert space H) by a nonlinear transformation, so that the hypersurface model in input space R^m corresponds to the hyperplane model in feature space H (support vector machine). Thus, the learning task of the classification problem can be solved by solving the linear support vector machine in the feature space.

(6) Kernel function

We define the kernel function as follows: let χ be an input space (subset or discrete set of R^m in the Euclidean space) and let H be a characteristic space (Hilbert space). If there exists a mapping from χ to H,

$$\varphi(x) : \chi \to H.$$

Let all $x, z \in \chi$, function $K(x, z)$ satisfy condition

$$K(x, z) = \varphi(x) \cdot \varphi(z).$$

$K(x, z)$ is the kernel function, and $\varphi(x)$ is the mapping function. In the formula, $\varphi(x) \cdot \varphi(z)$ is the inner product of $\varphi(x)$ and $\varphi(z)$. Under the condition that the kernels $K(x, z)$ are given as kernels, SVMs can be used to solve nonlinear classification problems by solving linear classification problems. Learning takes place implicitly in feature space, without explicitly defining feature space and mapping function. This method solves the nonlinear problem by ingeniously utilizing the linear classification learning method and the kernel function. In practical applications, it often depends on domain knowledge to select the kernel function directly. The validity of the kernel function selection needs to be verified by experiments.

The commonly used kernel functions include:

1. Polynomial kernel function

The corresponding support vector machine is a p-order polynomial classifier. In this case, the classification decision function becomes,

$$f(x) = \text{sign}\left(\sum_{i=1}^{N} \alpha_i^* y_i (x_i \cdot x + 1)^p + b^*\right).$$

2. Gaussian kernel function or radial basis function (RBF) is

$$K(x, z) = \exp\left(-\frac{||x - z||^2}{2\sigma^2}\right).$$

The corresponding support vector machine is a radial basis function classifier. In this case, the classification decision function is

$$f(x) = \text{sign}\left(\sum_{i=1}^{N} \alpha_i^* y_i \exp\left(-\frac{||x - z||^2}{2\sigma^2}\right) + b^*\right).$$

3. String kernel function

Kernel functions can be defined not only in Euclidean space, but also in the set of discrete data. For example, string validation is defined as a kernel function on a set of strings. String kernels are widely used in text classification, information retrieval, and bioinformatics.

Thus, we define the nonlinear support vector machine as: from the training set of nonlinear classification, through the maximization of kernel function and soft margin, or convex quadratic programming, the learning decision function is

$$f(x) = \text{sign}\left(\sum_{i=1}^{N} \alpha_i^* y_i K(x \cdot x_i) + b^*\right),$$

which is called the nonlinear support vector machine. $K(x, x_i)$ is a positive definite kernel function.

The learning algorithm of nonlinear support vector machine can be described as follows:

Input: Training data set $T = \{(x_1, y_1), (x_2, y_2), \ldots, (x_N, y_N)\}$, where $x_i \in X = R^*$, $y_i \in Y = \{+1, -1\}$, $i = 1, 2, \ldots, N$.

Output: Separate decision function.

Process:

1. Choosing appropriate kernel function $K(x, z)$ and appropriate parameter C to construct and solve the optimization problem,

$$\min_{a2} \sum_{i=1}^{N} \sum_{j=1}^{N} \alpha_i \alpha_j y_i y_j K\left(x_i \cdot x_j\right) - \sum_{i=1}^{N} \alpha_i$$

$$\text{s.t.} \ \sum_{i=1}^{N} \alpha_i y_i = 0,$$

$$C \geq \alpha_i \geq 0, \quad i = 1, 2, \ldots, N.$$

Obtaining the optimal solution $\alpha^* = (\alpha_1^*, \alpha_2^*, \ldots, \alpha_N^*)^T$.

2. Choose a positive component $C > \alpha_i^* > 0$ of α^*, and calculate

$$b^* = y_i - \sum_{i=1}^{N} \alpha_i^* y_i K(x_i \cdot x_j).$$

3. Constructing decision function is

$$f(x) = \text{sign}\left(\sum_{i=1}^{N} \alpha_i^* y_i K(x \cdot x_i) + b^*\right).$$

When $K(x, z)$ is a positive definite kernel function, the problem is a convex quadratic programming problem with global optimal solution, and many optimization algorithms can be used to solve this problem. However, when the training sample size is large, these algorithms often become very inefficient and cannot be used. So many fast algorithms have been put forward. Next, we introduce the Sequential Minimal Optimization (SMO) algorithm, which was proposed by Platt in 1998. Its characteristic is that the original quadratic programming problem is decomposed into quadratic programming sub-problems with only two variables, and the sub-problems are solved analytically until all variables satisfy the KKT condition bit. In this way, the optimal solution of the original quadratic programming problem is obtained by heuristic method, because the sub-problem has analytical solution, the algorithm is still efficient.

The dual problem of convex quadratic programming to be solved by SMO is as follows,

$$\min_{\alpha} \frac{1}{2} \sum_{i=1}^{N} \sum_{j=1}^{N} \alpha_i \alpha_j y_i y_j K(x_i \cdot x_j) - \sum_{i=1}^{N} \alpha_i,$$

$$\text{s.t.} \sum_{i=1}^{N} \alpha_i y_i = 0,$$

$$C \geq \alpha_i \geq 0, \quad i = 1, 2, \ldots, N.$$

In this problem, the variable is the Lagrange multiplier $\alpha^* = (\alpha_1^*, \alpha_2^*, \ldots, \alpha_N^*)^{\mathrm{T}}$. A variable α_i corresponds to a sample point (x_i, y_i); the total number of variables is equal to the training sample size N.

The SMO algorithm can be described as follows:

Input: Training data set $T = \{(x_1, y_1), (x_2, y_2), \ldots, (x_N, y_N)\}$, where $x_i \in X = R^*$, $y_i \in Y = \{+1, -1\}$, $i = 1, 2, \ldots, N$, precision ε.

Output: approximate solution of $\hat{\alpha}$.

Process:

1. Take the initial value of $\alpha^{(0)} = 0$, let $k = 0$.
2. The optimization variables $\alpha_1^{(k)}, \alpha_2^{(k)}$ are selected to solve the optimization problem of two variables analytically,

$$\min_{\alpha_1, \alpha_2} W(\alpha_1, \alpha_2) = \frac{1}{2} K_{11} \alpha_1^2 + \frac{1}{2} K_{22} \alpha_2^2 + y_1 y_2 K_{12} \alpha_1 \alpha_2$$

$$- (\alpha_1 + \alpha_2) + y_1 \alpha_1 \sum_{i=3}^{N} \alpha_i y_i K_{i1} + y_2 \alpha_2 \sum_{i=3}^{N} \alpha_i y_i K_{i2},$$

$$\text{s.t.} \, \alpha_1 y_1 + \alpha_2 y_2 = - \sum_{i=3}^{N} \alpha_i y_i = \zeta, C \geq \alpha_i \geq 0, \quad i = 1, 2, \ldots, N.$$

Obtaining the optimal solution $\alpha_1^{(k+1)}, \alpha_2^{(k+1)}$, and update α to $\alpha^{(k+1)}$.

3. If the shutdown condition is satisfied within the range of accuracy,

$$\sum_{i=1}^{N} \alpha_i y_i = 0,$$

$$C \geq \alpha_i \geq 0, i = 1, 2, \ldots, N,$$

$$y_i \cdot g(x_i) = \begin{cases} \geq 1, \{x_i | \alpha_i = 0\}, \\ = 1, \{x_i | 0 < \alpha_i < C\}, \\ \leq 1, \{x_i | \alpha_i = C\}, \end{cases}$$

where $g(x_i) = \sum_{j=1}^{N} \alpha_j y_j K(x_j \cdot x_i) + b$ goes to 4), otherwise $k = k + 1$ goes to 2).

4. Take $\hat{\alpha} = \alpha^{(k+1)}$.

4.1.7.2 Support Vector Classification Application

In the last section, we elaborated the principle of SVM on classification (we call it Support Vector Classification). Here, we refer to Scikit-learning project and give several different examples of SVC application [7]. Specifically, we use linear, RBF and polynomial to verify the classification training and prediction of Iris data set.

4.1.7.3 Support Vector Regression

Support vector classification can be extended to solve regression problems. This method is called support vector regression. The model generated by support vector classification only depends on a subset of data, because the cost function of building the model does not care about the training points beyond the boundary. Similarly, the model generated by support vector regression only depends on a subset of training data, because the cost function of building the model ignores any training data close to the model prediction. Like classification classes, the fitting method will be used as parameter vectors x and y. Only in this case, y expects to have floating point values instead of integer values.

For the definition of SVR, we give a training data set

$$T = \{(x_1, y_1), (x_2, y_2), \dots, (x_N, y_N)\},$$

Where $x_i \in X = R^*, y_i \in Y = \{+1, -1\}, i = 1, 2, \dots, N$; then the learning problem of SVR becomes the original problem of learning as follows,

$$\min_{w,b,\xi,\xi^*} \frac{1}{2} w^{\mathrm{T}} w + C \sum_{i=1}^{N} \xi_i + \xi_i^*,$$

$$\text{s.t.} y_i - w^{\mathrm{T}} \phi(x_i) - b \leq \varepsilon + \xi_i, \quad i = 1, 2, \dots, N,$$

$$w^{\mathrm{T}} \phi(x_i) + b - y_i \leq \varepsilon + \xi_i^*, \quad i = 1, 2, \dots, N,$$

$$\xi_i, \xi_i^* \geq 0, \quad i = 1, 2, \ldots, N.$$

Its dual problem can be transformed into,

$$\min_{\alpha,a^*} \frac{1}{2}(\alpha - \alpha^*)^{\mathrm{T}} Q(\alpha - \alpha^*) + \varepsilon e^{\mathrm{T}}(\alpha + \alpha^*) - y^{\mathrm{T}}(\alpha - \alpha^*)$$

$$\text{s.t. } e^{\mathrm{T}}(\alpha - \alpha^*) = 0,$$

$$C \geq \alpha_i, \alpha_i^* \geq 0, \quad i = 1, 2, \ldots, N$$

Where e is a vector of all $1, C \geq 0$ is the upper line, Q is a semi-positive definite matrix of $N*N$, and the kernel $Q_{ij} \equiv K(x_i, y_i) = (x_i)^{\mathrm{T}}(x_j)$. The training vector is implicitly mapped to a higher dimensional (possibly infinite dimension) dimension space by the function.

The decision function is

$$\sum_{i=1}^{n}(\alpha - \alpha^*)K(x_i, x) + \rho.$$

4.1.7.4 Pros and Cons

(1) Pros

The advantages are:

1. Nonlinear mapping is the theoretical basis of SVM method. SVM uses inner product kernel function to replace the nonlinear mapping to high-dimensional space.
2. The optimal hyperplane for feature space partition is the goal of SVM, and the idea of maximizing classification margin is the kernel of SVM.
3. Support vector is the training result of SVM. Support vector plays a decisive role in SVM classification decision.
4. SVM is a novel small sample learning method with solid theoretical basis. It basically does not involve probability measure and law of large numbers, so it is different from the existing statistical methods. Essentially, it avoids the traditional process from induction to deduction, realizes efficient "transduction reasoning" from training samples to prediction samples, and greatly simplifies the usual classification and regression problems.
5. The final decision function of SVM is determined only by a few support vectors. The complexity of calculation depends on the number of support vectors, not the dimension of sample space, which avoids the "dimension disaster" in a sense.

6. A small number of support vectors determine the final result, which not only helps us grasp the key samples and "eliminate" a large number of redundant samples, but also dooms the method to be not only simple in algorithm and has good robustness. This "robustness" is mainly embodied in: ① adding or deleting non-support vector samples has no effect on the model; ② support vector samples have certain robustness; ③ in some successful applications, the SVM method is insensitive to the selection of the kernel.

(2) **Cons**

1. SVM algorithm is difficult to implement for large-scale training samples.

Because SVM solves support vector by means of quadratic programming, solving quadratic programming will involve the calculation of m-order matrix (m is the number of samples). When the number of M is large, the storage and calculation of the matrix will consume a lot of machine memory and computing time. The main improvements to the above problems are the SMO algorithm of J. Platt, SVM of T. Joachims, PCGC of C. J. C. Burges, and SOR algorithm of O. L. Mangasarian.

2. It is difficult to solve multi-classification problems with SVM.

Classical SVM algorithm only gives two classes of classification algorithm, but in the practical application of data mining, it is generally necessary to solve the problem of multi-class classification. It can be solved by combining multiple two-class support vector machines. There are mainly one-to-many combination mode, one-to-one combination mode, and SVM decision tree, and then it is solved by constructing the combination of multiple classifiers. The main principle is to overcome the inherent shortcomings of SVM and combine the advantages of other algorithms to solve the classification accuracy of multi-class problems.

4.1.8 Ensemble Methods

The ensemble algorithm uses a variety of learning algorithms to achieve better prediction performance than any single ensemble learning algorithm. Machine learning sets contain only a set of specific finite sets of alternative models, but usually allow for more flexible structures in these alternative algorithms. Researchers have made great efforts on which learning algorithms to combine and how to combine. It is a very powerful technology, and it is very popular. The aim of ensemble algorithm is to construct several basic estimator predictions by using a given learning algorithm in order to improve the universality/robustness of a single estimator.

The most commonly used integrated learning methods include bagging and boosting. Simply put, in bagging method, individual learners do not depend on each other, and they can be generated by parallel random sampling of samples at the same time. It generates multiple independent models for random sampling of samples and feature attributes and then averages the predicted values of all models in order to reduce variance. Typical methods include bagging and random forests. In boosting method, there is a strong dependence between individual learners, which generates a new model based on the training results (errors) of the previous model, so it must be serialized. Boosting method is used to fit the error of the previous model in order to reduce the deviation. Typical boosting methods include Adaboost and GBDT. We will introduce it in more detail below.

1. **Bagging (bootstrapped aggregation)**

Bagging is the abbreviation of Bootstrap AG GregatING. Bagging is based on bootstrap sampling. Given a data set containing m samples, a sample is randomly taken out and put into the sample set, and then the sample is put back into the initial data set, so that the sample may still be selected in the next sampling. In this way, after M times of random sampling operation, we get a sample set with m samples. Some samples in the initial training set appear many times in the resampling set, while others never appear. T sample sets containing m training samples are sampled, and then a basic learner is trained based on each sample set, and these basic learners are combined. Bagging usually uses simple voting method for divided tasks and simple averaging method for regression tasks. Bagging can be used for multi-classification, regression, and other tasks without modification. The self-help sampling process also brings another advantage to Bagging: since each base learner uses only about 63.2% of the original training set, the remaining 36.8% of the samples can be used as validation sets to "out-of-bag" estimation of generalization performance.

2. **Random forest**

Random forest, as its name implies, is a forest composed of many decision trees. Random means that every tree has no connection and is independent. Random Forest actually adds a condition to bagging. It also sampled repeatedly according to the bagging method, but the number of samples sampled was equal to the total number of samples (n). But not all features are used in training trees. Suppose we have a total of M features. Each time we train a tree, we randomly select $m(<<M)$ features for training. Trees in random forests need not be pruned. Because of the sample extraction, feature extraction has guaranteed randomness, greatly reducing the possibility of overfitting.

The methods for generating random forests can be described as follows:

1. Selecting n data from training data as input of training data, generally n is much smaller than the whole training data N, which will result in some data cannot be accessed. This part is called out-of-pocket data and can be used for error estimation.
2. After selecting the input training data, it is necessary to construct a decision tree. The specific method is to select m features from the whole feature set M for each split node. In general, m is much smaller than M.
3. In the process of constructing each decision tree, the decision tree is constructed by choosing the smallest Gini index and choosing the splitting nodes. The other nodes of the decision tree are constructed with the same splitting rules until all training samples of the node belong to the same class or reach the maximum depth of the tree.

4. Repeat steps 2 and 3 several times, each time the input data corresponds to a decision tree, so that a random forest can be obtained, which can be used to make decisions on the prediction data.
5. The input training data are well selected, and multiple decision trees are constructed. The prediction data are processed, such as input a data to be predicted, then multiple decision trees make decisions simultaneously, and finally use the majority voting method to make category decisions.

In a word, random forest is an excellent model. It has a high efficiency for multi-dimensional feature data set classification and can also make the choice of feature importance. The operation efficiency and accuracy rate are high, and the implementation is relatively simple. However, in the case of large data noise, it will be overfitting. The disadvantage of overfitting is fatal for random forest. For data with different attributes, more attributes with different values will have a greater impact on random forests, so the attribute weights produced by random forests on this data are not credible.

In summary, the random forest and bagging mainly have two main differences: (1) Random forest is selected as many times as the number of input samples (may be one sample will be selected multiple times, and some samples will not be picking), while bagging generally selects fewer samples than the number of input samples; (2) Bagging uses all features to get the classifier, and random forest needs to select some of the features to train the classifier; the random forest is generally better than bagging.

3. Adaboost

Boosting method is a method to improve the accuracy of weak classification algorithm. This method constructs a series of predictive functions and then combines them into a predictive function in a certain way. It is a framework algorithm, which obtains the sample subset through the operation of the sample set, and then trains the weak classification algorithm to generate a series of base classifiers on the sample subset. It can be used to improve the recognition rate of other weak classification algorithms, i.e., putting other weak classification algorithms in the boosting framework as the base classification algorithm. Through the operation of the boosting framework on the training sample set, different subsets of training samples are obtained, and the sample subset is used to train and generate the base classifier; each sample set is generated by the base classification algorithm on the sample set, so that N base classifiers can be generated after a given number of training rounds n. Then boosting framework algorithm weights the N base classifiers and produces a final result classifier. In the N base classifiers, the recognition rate of each single classifier is not necessarily very high, but their combined results have a high recognition rate, which improves the recognition rate of the weak classification algorithm. When generating a single base classifier, the same classification algorithm or different classification algorithms can be used. These algorithms are generally unstable weak classification algorithms, such as neural network (BP), decision tree (C4.5), and so on.

Adaboost is a representative algorithm in boosting. The basic idea is to construct a classifier through the distribution of training data, and then calculate the weight of the weak classifier through the error rate. By updating the distribution of training data, it iterates until the number of iterations or the loss function is less than a certain threshold. Specifically, at the beginning of the algorithm, each sample is assigned a weight value. At the beginning, everyone is equally important. At the end of each step, we increase the weights of the wrong points and reduce the weights of the points, so that if some points are always misjudged, they will be "seriously concerned" and given a high weight. Then we do N iterations (designated by users) to get N simple learners, and then we combine them (such as weighting them or voting them) to get a final model.

Adaboost's algorithm flow:

Suppose the training data set is $T = \{(x_1, y_1), (x_2, y_2), (x_3, y_3), (x_4, y_4), (x_5, y_5)\}$, where $y_i = \{-1, +1\}$.

1. Initialization of training data distribution

The weight distribution of training data is $D = \{w_{11}, w_{12}, w_{13}, w_{14}, w_{15}\}$, where $w_{1i} = 1/N$, that is to say, the average distribution.

2. Selecting basic classifier

The simplest linear classifier $y = aX + b$ is chosen here. After the classifier is selected, the parameters can be obtained by minimizing the classification error.

3. Calculating the coefficients of classifiers and updating data weights

The error rate can also be calculated as e_1. At the same time, the coefficients of the classifier can be calculated. The basic formula for calculating the coefficients given by Adaboost is as follows,

$$a_m = \frac{1}{2} \log \frac{1 - \exp_m}{\exp_m}.$$

Then the weight distribution of training data is updated,

$$D_{m+1} = (w_{m+1,1}, \ldots, w_{m+1,i}, \ldots, w_{m+1,N}),$$

$$w_{m+1,i} = \frac{w_{mi}}{Z_m} \exp(-\alpha_m y_i G_m(x_i)), \quad i = 1, 2, \ldots, N.$$

Here, Z_m is the normalization factor,

$$Z_m = \sum_{i=1}^{N} w_{mi} \exp(-\alpha_m y_i G_m(x_i)).$$

It makes D_{m+1} a probability distribution.

4. Combination of classifiers is,

$$f(x) = \sum_{m=1}^{M} \alpha_m G_m(x).$$

Of course, this combination method is based on the coefficients of the classifier, and the coefficients of the classifier are calculated according to the error rate, so the final impact of Adaboot is how to use the error rate, as well as the calculation coefficient of training data update weight.

In the specific application of Adaboost, there are some parameters that can be adjusted and optimized. We can optimize the algorithm from these aspects: how to select weak classifier; how to calculate the coefficient of classifier with better experimental error rate; how to calculate the weight distribution of training data better; how to combine weak classifiers; how much are the threshold of the loss function and the number of iterations.

4. Gradient boosted regression trees (GBRT)

Gradient tree boosting is also called (gradient boosted regression trees (GBRT) or gradient boosted decision trees (GBDT), boosted tree and so on. Boosted tree is actually a framework, which can be embedded in many different algorithms; there are many different varieties. The difference between gradient boost and traditional boost is that each calculation is to reduce the last residual, and in order to eliminate the residual, we can build a new model in the gradient direction of residual reduction. Therefore, in gradient boost, each new model is built to reduce the residual gradient direction of the previous model, which is quite different from the traditional boost in weighting the correct and wrong samples. GBRT can be used for regression and classification. The only difference between GBRT and GBRT is that the loss function is different.

GBRT considers the form of additive models,

$$f(x) = \sum_{m=1}^{M} \gamma_m h_m(x).$$

Among them, $h_m(x)$ denotes the basis function, which is also called weak learner in boosting algorithm. GBRT uses a fixed number of decision trees as weak learners. Similar to other Boosting algorithms, GBRT establishes an additive model in the form of forward stage wise,

$$F_m(x) = F_{m-1}(x) + \gamma_m h_m(x).$$

In each stage, given the existing model F_{m-1} and its fit $F_{m-1}(x_i)$, the decision tree basis function $h_m(x)$ is used as the minimization loss function L,

$$F_m(x) = F_{m-1}(x) + \arg\min_h \sum_{i=1}^{n} L(y_i, F_{m-1}(x_i) + h(x)).$$

The initial value F_0 of the model needs to be set according to the problem. For example, least squares regression usually chooses the mean of the target value. Initial models can also be specified by init parameters. The objects transferred must be fitted and predicted. GBRT attempts to solve this minimization problem numerically by the steepest descent method, steepest descent: the steepest descent direction is the negative gradient of the loss function evaluated in the current model, which can be calculated by any differentiable loss function,

$$F_m(x) = F_{m-1}(x) - \gamma_m \sum_{i=1}^{n} \nabla_F L(y_i, F_{m-1}(x_i)).$$

Use line search to select the step size γ_m,

$$\gamma_m = \arg\min_\gamma \sum_{i=1}^{n} L(y_i, F_{m-1}(x_i) - \gamma \frac{\partial L(y_i, F_{m-1}(x_i))}{\partial F_{m-1}(x_i)}).$$

GBRT is one of the most commonly used algorithms. Because it works well and is insensitive to input requirements, it is often one of the necessary tools from statisticians to data scientists. At the same time, because of its good effect and low computational complexity, it has also been widely used in industry. For example, GBRT derives a very popular algorithm called xgboost, which is a tool for large-scale parallel boosted tree. It is the fastest and best open source boosted tree toolkit at present, more than ten times faster than common toolkits. In data science, a large number of kaggle players choose it for data mining competitions, including more than two kaggle competitions to win the championship. In terms of industrial scale, the distributed version of xgboost has a wide range of portability, supports running on multiple platforms, and retains various optimizations of single-machine parallel version, which makes it a good solution to the problem of industrial scale.

In short, the main difference between bagging and boosting is the different sampling methods. Bagging uses uniform sampling, while boosting uses error rate to sample, so boosting's classification accuracy is better than bagging's. The selection of bagging's training set is random, and each training set is independent of each other, while the selection of boosting's training set is related to the learning results of the previous rounds; Bagging's prediction functions have no weight, while boosting is weighted; Bagging's prediction functions can be generated in parallel, while boosting's prediction functions can only be generated sequentially. For extremely time-consuming learning methods such as neural networks, bagging can save a lot of time and overhead through parallel training. Both bagging and boosting can effectively improve the accuracy of classification. In most data sets, boosting is more

Table 4.2 Comparison of bagging and boosting algorithms

	Bagging	Boosting
Sampling mode	Bagging uses uniform sampling	Boosting samples based on error rate
Accuracy	In contrast, it is lower	High
Training set selection	Random, independent training sets	The selection of each training set is related to the learning results of the previous rounds
Weight of predictive function	Each prediction function has no weight	Boost has weight
Function generation order	Parallel generation	Sequential generation

accurate than bagging. In some data sets, boosting causes degradation—overfitting. The comparison between the two is summarized in Table 4.2.

4.2 Unsupervised/Semi-supervised Learning Models

4.2.1 Dimensionality Reduction Algorithms

Dimensionality reduction models seek and utilize the intrinsic structure of data in order to use less information to summarize or describe data. Dimension reduction is the process of reducing the number of random variables considered by obtaining a set of main variables. It can be divided into feature selection and feature extraction. After visualizing or simplifying the high-dimensional data, the regression and classification methods used in supervised learning have a good effect. We divide Jiangwei method into linear dimension reduction method and nonlinear dimension reduction method.

(1) **Linear dimensionality reduction algorithms**
1. **Principal component analysis (PCA)**

   PCA is a statistical method. By orthogonal transformation, a group of variables that may be correlated are transformed into a group of variables that are linearly unrelated. The transformed variables are called principal components. The idea is to study how to reveal the internal structure of multiple variables through a few principal components, that is, to derive a few principal components from the original variables, so that they retain as much information as possible about the original variables and are not related to each other. Generally, the mathematic treatment is to combine the original P indices linearly as a new comprehensive index.

   Suppose we have a set of training sets $\{x^{(i)}: i = 1, 2, \ldots, m\}$, contains m training samples, each training sample $x^{(i)} \in R^n$, where $(n \ll m)$, each n-dimensional training sample means that there are n attributes. Generally speaking, many of these attributes are related, that is, many attributes are redundant, which provides the possibility for feature dimensionality reduction. The key is how to determine the redundant attributes and how to reduce the dimension.

PCA provides a solution to this problem. Before doing PCA, we need to do the following data preprocessing:

1. Find the mean vector of training set, $\mu = \frac{1}{m} \sum_{i=1}^{m} x^{(i)}$.

2. Subtract the mean vector from each training sample, $x^{(i)} = x^{(i)} - \mu$.

3. Calculate the variance of the transformed training set, $\sigma_j^2 = \frac{1}{m} \sum_i \left(x_j^{(i)} \right)^2$.

4. Substitute the sample of training set as follows, $x_j^{(i)} = x_j^{(i)} / \sigma_j$.

The first and second steps above ensure that the mean value of the training set is 0, and the third and fourth steps ensure that the variance of the training set is 1, so that the different attributes in the training sample are transformed to the same scale. Given a unit vector μ and a point x, the projection length from the point x to the unit vector is $x^T \mu$. If $x^{(i)}$ is a sample in the training set, its projection length on μ is the distance from $x^T \mu$ to the origin. Therefore, in order to maximize the variance between these projections, we hope to find the unit vector μ satisfying the following expressions,

$$\frac{1}{m} \sum_{i=1}^{m} \left(\left(x^{(i)} \right)^T \mu \right)^2 = \frac{1}{m} \sum_{i=1}^{m} \mu^T x^{(i)} \left(x^{(i)} \right)^T \mu = \mu^T \left(\frac{1}{m} \sum_{i=1}^{m} x^{(i)} \left(x^{(i)} \right)^T \right) \mu.$$

Because μ is a unit vector, so $|\mu|^2 = 1$, the expression in parentheses above is the covariance matrix $\left(\Sigma = \frac{1}{m} \sum_{i=1}^{m} x^{(i)} \left(x^{(i)} \right)^T \right)$ whose mean value is 0. In order to maximize the objective function, μ should take the eigenvector corresponding to the largest eigenvalue of Σ.

In short, we should take the principal eigenvector of Σ, if we want to map the original data space to a low-dimensional subspace, we can choose the first k eigenvectors of as the basis vectors of the subspace, then the k eigenvectors of μ_1, μ_2, ..., μ_k constitutes the base vector of the new space. Then we can map the original training sample $x^{(i)}$ to a new feature space,

$$y^{(i)} = \begin{bmatrix} \mu_1^T x^{(i)} \\ \cdots \\ \mu_k^T x^{(i)} \end{bmatrix} R^k.$$

Therefore, although $x^{(i)}$ is an n-dimensional vector, $y^{(i)}$ becomes a lower-dimensional vector, so PCA is a dimension reduction algorithm, in which the eigenvectors μ_1, μ_2, ..., μ_k are called the first k principal component of the training set.

2. **Linear discriminant analysis (LDA)**

LDA and PCA are commonly used dimension reduction methods. The difference between them is:

1. Different starting ideas. PCA mainly finds a better projection method from the angle of covariance of features, that is, to select the direction in which the sample point projection has the largest variance; while LDA considers the classification label information more, and seeks to maximize the distance between data points of different categories and minimize the distance between data points of the same category after projection, that is, to select the direction with the best classification performance.
2. Different learning modes. PCA belongs to unsupervised learning, so it is only a part of data processing in most scenarios and needs to be combined with other algorithms, such as PCA and clustering, discriminant analysis, regression analysis, etc. LDA is a supervised learning method, which can be used not only to reduce dimension, but also to predict, so it can be combined with other models as well used independently.
3. The number of available dimensions is different after dimension reduction. LDA can generate $(K - 1)$-dimensional subspace at most after dimensionality reduction (number of classification labels $= K$), so LDA has nothing to do with the number of original dimensions, only the number of data labels; PCA has n dimensions at most, that is to say, it can choose all available dimensions at most.

From the point of view of direct visualization, taking two-dimensional data dimensionality reduction as an example, the LDA is as in Fig. 4.2.

On the left side of the picture above is the idea of PCA dimensionality reduction. All it does is to map the whole set of data onto the coordinate axis which is most convenient to represent the set of data. No classification information is used in the mapping. Therefore, although the data after PCA are more convenient to represent (reduce dimension and keep original information to the maximum), it may become

Fig. 4.2 LDA algorithm

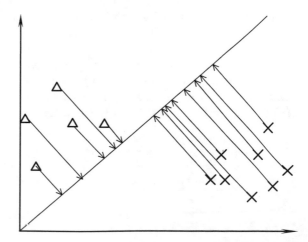

more difficult to classify. On the right side of the picture above is the idea of LDA dimensionality reduction. It can be seen that LDA makes full use of the classification information of data and maps two sets of data to another coordinate axis, which makes the data more easily distinguishable (in the lower dimension it can distinguish and reduce the computational complexity).

In the theory of classifier, Bayesian classifier is the best classifier. In order to get the best classifier, we need to know the posterior probability $P(C_k|x)$ of the classes. It is assumed that $f_k(x)$ is the class conditional probability density function of class C_k and that π_k is the prior probability of class C_k. There is no doubt that there is $\sum_k \pi_k = 1$. According to Bayesian theory,

$$P(C_k|x) = \frac{f_k(x)\pi_k}{\sum_{l=1}^{K} f_l(x)\pi_l}.$$

Because π_k is almost known, the most important thing for Bayesian formulas is the conditional probability density function $f_k(x)$. The main reason why many algorithms are different is that the assumptions about the parameter form of such conditional probability density function are different, such as:

Linear discriminant analysis (LDA) assumes that $f_k(x)$ is a Gauss distribution with different mean and the same variance.

Quadratic discriminant analysis (QDA) assumes that $f_k(x)$ is a Gaussian distribution with different mean and variance.

The Gaussian mixture model (GMM) assumes that $f_k(x)$ is a combination of different Gaussian distributions.

As mentioned earlier, LDA assumes that $f_k(x)$ is a Gauss distribution with different mean and the same variance, so its conditional probability density function can be written as follows,

$$f_k(x) = \frac{1}{(2\pi)^{p/2}|\Sigma|^{1/2}} \exp\left(-\frac{1}{2}(x - \mu_k)^T \sum{}^{-1}(x\mu_k)\right).$$

Here, the dimension of feature x is p-dimension, the mean value of category C_k is μ_k, and the variance of all categories is Σ.

As mentioned earlier, after adding a monotone function $f(\cdot)$ to the discriminant function $\delta_k(x)$ or the posterior probability function $P(C_k|x)$, a linear classifier can obtain that the transformed function is a linear function of x, and the obtained linear function is the decision surface. The monotone transformation function $f(\cdot)$ used by LDA is logit function: $\log[p/(1-p)]$. For binary classification problems, there is,

$$\log \frac{P(C_1|x)}{P(C_2|x)} = \log \frac{f_1(x)}{f_2(x)} + \log \frac{\pi_1}{\pi_2}$$
$$= x^T\Sigma^{-1}(\mu_1 - \mu_2) - \frac{1}{2}(\mu_1 + \mu_2)^T\Sigma^{-1}(\mu_1 - \mu_2) + \log \frac{\pi_1}{\pi_2}.$$

It can be seen that the decision-making plane is a plane.

According to the above formula, it is also easy to get the decision function of LDA as follows,

$$\delta_k(x) = x^{\mathrm{T}}\Sigma^{-1}\mu_k - \frac{1}{2}\mu_k^{\mathrm{T}}\Sigma^{-1}\mu_k + \log\pi_k.$$

The parameters are estimated from the data: $\pi_k = N_k/N$, N_k is the sample number of category C_k and N is the total sample number. $\mu_k = \frac{1}{N_k}\sum_{xC_k} x$ is the sample mean of category C_k.

$$\Sigma = \frac{1}{N-K}\sum_{k=1}^{K}\sum_{xC_k}(x-\mu_k)(x-\mu_k)^{\mathrm{T}}.$$

We extend the quadratic discriminant analysis QDA. The quadratic discriminant function assumes that $f_k(x)$ is of different mean and variance. Compared with LDA, because Σ_k is different, its quadratic term exists, so its decision-making surface is as follows,

$$\log\frac{P(C_1|x)}{P(C_2|x)} = \log\frac{f_1(x)}{f_2(x)} + \log\frac{\pi_1}{\pi_2} = x^{\mathrm{T}}(\Sigma_1^{-1} - \Sigma_2^{-1})x$$
$$+ x^{\mathrm{T}}\Sigma^{-1}(\mu_1 - \mu_2) - \frac{1}{2}(\mu_1 + \mu_2)^{\mathrm{T}}\Sigma^{-1}(\mu_1 - \mu_2) + \log\frac{\pi_1}{\pi_2} - \frac{1}{2}\log\frac{|\Sigma_1|}{|\Sigma_2|}.$$

The corresponding discriminant function is

$$\delta_k(x) = -\frac{1}{2}(x-\mu_k)^{\mathrm{T}}\Sigma^{-1}(x-\mu_k) - \frac{1}{2}\log|\Sigma_k| + \log\pi_k.$$

After understanding the principles of LDA and QDA in classification, let us look at the use of LDA in dimension reduction. In order to understand how LDA reduces dimension, we first reconstruct the above classification rules. K is the total number of target classes, and it is known that all LDA classes have the same estimated covariance, so we readjust the data so that the covariance is uniform,

$$X* = D^{-1/2}U^{\mathrm{T}}X \quad \text{with} \sum = UDU^{\mathrm{T}}.$$

Then it can be proved that the classification of data points after adjustment is equivalent to finding the estimated class mean closest to the data points in Euclidean distance. However, this can also be done by projecting all classes $\overset{*}{\mu} k$ into $K-1$ affine subspaces H_K. This shows that there is a dimension reduction process in LDA classifier by linear projection to $(K-1)$-dimensional space. Further, we can reduce the dimension to L more by maximizing variance $\overset{*}{\mu} k$ to project H_L into linear subspace $\overset{*}{\mu} k$.

(2) **Nonlinear dimensionality reduction algorithms**
1. **Multidimensional scaling (MDS)**

Multidimensional Scaling (MDS) attempts to find a low-dimensional representation of data, where the distance corresponds to the distance in the original high-dimensional space. Generally speaking, it is used to analyze similarity or dissimilarity data. MDS attempts to model similarity or dissimilarity data as distances in geometric space. Data can be the ratio of similarities between objects, the frequency of interactions between molecules, or trade indices between countries.

There are two types of MDS algorithms: metric and non-metric. In metric MDS, the input similarity matrix is derived from the measurement of the output distance (so triangular inequality is considered), and then set to be as close as possible to the similarity or dissimilarity data. In the non-metric version, the algorithm will try to maintain the order of distance and therefore seek the monotonous relationship between distance and similarity/dissimilarity in embedded space.

2. **Isomap**

Isomap algorithm is an algorithm derived from MDS algorithm. MDS algorithm keeps the distance between samples unchanged after dimension reduction. Isomap algorithm introduces neighborhood graph. Samples are only connected with adjacent samples, and the distance between them can be calculated directly. The distant points can be calculated by the minimum path. On this basis, dimension reduction and distance preservation can be carried out.

The calculation process is as follows:

1. Set the number of neighborhood points and calculate the adjacent distance matrix. The distance outside the neighborhood is set to infinity.
2. Find the minimum path between each pair of points and transform the adjacency matrix into the minimum path matrix.
3. Input the MDS algorithm and get the result, which is the result of Isomap algorithm.

Floyd algorithm is used in the minimum path: input adjacency matrix; in the adjacency matrix, except for the neighborhood points, the other distances are infinite, and output the complete distance matrix.

3. **Locally-linear embedding (LLE)**

Locally linear embedding, LLE, is a kind of nonlinear dimension reduction algorithm. It uses the local symmetry of linear reconstruction to find out the nonlinear structure in high-dimensional data space, and maps high-dimensional data points to corresponding data points in low-dimensional space while maintaining the close position relationship of each data point. It can keep the original popular structure of reduced-dimensional data.

The calculation process is as follows:

1. Find k-nearest neighbors of each sample point.
2. Calculate the local weight matrix of each sample point from its nearest neighbors.
3. The output value of the sample points is calculated by reconstructing the weight matrix of the neighboring points and the sample points.

For each block X_1 in image X, the first step is to calculate k nearest neighbors of each sample point X_i, that is, to find the nearest neighbor set of the sample point, where the calculation can be obtained by Euclidean distance method; the second step is to calculate the local reconstruction weight matrix W of the sample point to minimize the error of the reconstruction block X_1; the third step is to map all the sample points into low-dimensional space to achieve dimensionality reduction.

4. **Spectral embedding**

Spectral embedding is a method for calculating nonlinear embedding. The algorithm uses the spectral decomposition of the Laplacian operator to find the low-dimensional representation of the data through Laplacian feature mapping, and the resulting graph can be considered as the discrete approximation of the low-dimensional manifold in the high-dimensional space. The minimization of the cost function based on the graph ensures that the points close to each other on manifolds are close to each other in low-dimensional space, thus maintaining local distance.

The calculation process is as follows:

1. Construction of weighted graph. The original input data are converted to a graphical representation using an association (adjacency) matrix representation.
2. Tulaplas structure. The non-standard Tulaplas algorithm is constructed and normalized.
3. Partial eigenvalue decomposition. Eigenvalue decomposition is accomplished on the Tulaplacian operator.

4. ***t*-distributed stochastic neighbor embedding (*t*-SNE)**

t-SNE (tSNE) converts the affinity of data points into probability. Affinity in primitive space is expressed by Gauss joint probability, and affinity in embedded space is expressed by student's t distribution. This makes t-SNE particularly sensitive to local structures and has some other advantages compared with existing technologies: displaying structures in multiple scales on a single map; revealing data in multiple manifolds or clusters; and reducing the tendency to focus on central aggregation points.

Although Isomap, LLE and variants are best suited to expand a single continuous low-dimensional manifold, t-SNE will focus on the local structure of the data and tend to extract clustered local sample groups. This ability of grouping samples based on local structure may be beneficial to simultaneously visually unravel data sets containing multiple manifolds in digital data sets.

4.2.2 Clustering Algorithms

Each clustering algorithm has two variables. A class: it realizes learning clustering on training set by fitting method; a function: returning a list of integer labels corresponding to different clustering according to given training set. There are many clustering methods, including K-means, affinity propagation, mean shift, spectral clustering, Ward hierarchical clustering, DBSCAN, Gaussian mixtures, birch. Here we give a brief introduction (see Table 4.3).

(1) K-means

The flow chart of the algorithm is as follows:

1. Select the initial value of μ_k for K centers. This process usually has some heuristic selection methods for specific problems, or in most cases uses the random selection method. Because K-means cannot guarantee the global optimum, but whether

Table 4.3 Overview of clustering methods

Method name	Parameter	Scalability	Use case	Geometry (metric used)
K-means	Number of clusters	Very large N samples, medium K clusters with MiniBatch code	General-purpose, even cluster size, flat geometry, not too many clusters	Distances between points
Affinity propagation	Damping, sample preference	Not scalable with N samples	Many clusters, uneven cluster size, non-flat geometry	Graph distance (e.g., nearest-neighbor graph)
Ward hierarchical clustering	Number of clusters	Large N samples and K clusters	Many clusters, possibly connectivity constrains	Distances between points
Agglomerative clustering	Number of clusters, linkage type, distance	Large N samples and K clusters	Many clusters, possibly connectivity constrains, non-Euclidean distance	Any pairwise distance
Dbscan	Neighborhood size	Very large N samples, medium K clusters	Non-flat geometry, uneven cluster sizes	Distances between points
Gaussian mixtures	Many	Not scalable	Flat geometry, good for density estimation	Mahalanobis distances to centers
Birch	Branching factor, threshold, optional global clusterer	Large N samples and K clusters	Large data set, outlier removal, data reduction	Euclidean distance between points

it converges to the global optimum has a great relationship with the selection of initial value, so sometimes we choose the initial value to run K-means many times and take the best one.

2. Categorize each data point into the cluster represented by the nearest central point.
3. Whenever a point is assigned to a certain category, the new central point of the cluster is recalculated with a formula,

$$\mu_k = \frac{1}{N_k} \sum_{j \in \text{cluster}_k} x_j.$$

4. Repeat the second step until the maximum number of steps is iterated or the difference between the values of J before and after iteration is less than a threshold.

(1) Affinity propagation (AP)

Affinity propagation creates clusters by sending messages between sample pairs until they converge. A few samples are then used to describe the data set, and these samples are identified as the most representative of other samples. Messages sent between them indicate that one sample is suitable as an example of another sample, which is updated based on values from other pairs. This update iteration occurs until convergence, at which time the final example is selected and the final clustering is given.

Affinity propagation can be interesting as it chooses the number of clusters based on the data provided. For this purpose, the two important parameters are the preference, which controls how many exemplars are used, and the damping factor which damps the responsibility and availability messages to avoid numerical oscillations when updating these messages.

Algorithmic description:

Messages sent between points fall into one of two categories. The first is responsibility $r(i, k)$, which indicates that sample K should be cumulative evidence of exemplar of sample i; the second is availability $\alpha(i, k)$, which indicates that sample I should select sample K as cumulative evidence of exemplar, and consider the value of all other samples when sample K is used as exemplar. In this way, when exemplars (1) are sufficiently similar to many samples and (2) are selected by many samples to represent themselves, exemplars are selected by samples.

More formally, the responsibility of a sample K to be the exemplar of sample I is given by

$$r(i, k) \leftarrow s(i, k) - \max[a(i, k') + s(i, k') \forall k' \neq k],$$

where $s(i, k)$ is the similarity between sample i and k. The availability of sample K to be the exemplar of sample I is given by

$$a(i, k) \leftarrow \min[0, r(k, k) + \sum_{i's.t.i'\notin\{i,k\}} r(i', k)].$$

Firstly, all the values of r and α are set to 0, and then each iteration is computed until convergence. As mentioned above, in order to avoid numerical oscillation when updating the message, the attenuation factor is introduced into the iteration process,

$$r_{t+1}(i, k) = \lambda \cdot r_t(i, k) + (1 - \lambda) \cdot r_{t+1}(i, k),$$

$$a_{t+1}(i, k) = \lambda \cdot a_t(i, k) + (1 - \lambda) \cdot a_{t+1}(i, k),$$

where t represents the number of iterations.

(3) **Mean shift**

Average shift clustering aims at finding speckles in smooth density samples. This is a centroid-based algorithm, which updates the candidate values of the centroid of moments as the average of points in a given region. Then these candidates are filtered in the post-processing stage to eliminate the approximate duplicate terms to form the final centroid set.

Given the candidate centroid, for iteration t, the candidate is updated according to the following equation,

$$x_i^{t+1} = x_i^t + m(x_i^t),$$

where $N(x_i)$ is the neighborhood of the sample x_i around a given distance, and m is the mean shift vector, which is calculated from each centroid of the region of maximum increase pointing to the point density. The centroid can be effectively updated to the average of the samples in its neighborhood by the following formula,

$$m(x_i) = \frac{\sum_{x_j \in N(x_i)} K(x_j - x_i)x_j}{\sum_{x_j \in N(x_i)} K(x_j - x_i)}.$$

The algorithm is not highly scalable because it requires multiple nearest neighbor searches during the execution of the algorithm. The algorithm guarantees convergence, but when the centroid changes very little, the algorithm stops iterating.

(4) **Spectral clustering**

Spectral clustering embeds affinity matrix between samples in low dimension and then uses K-means in low dimension space. If the affinity matrix is sparse and pyamg module is installed, it is particularly effective. Spectral clustering requires specifying the number of clusters. It applies to a few clusters, but is not recommended for multiple clusters.

For two clusters, it solves the convex relaxation of the normalized cut problem on similar graphs: cut the graph into two halves so that the weight of the cut edges is smaller than that of the edges in each cluster. This standard is particularly interesting when dealing with images: the vertex of a graph is a pixel, and the edge of a similarity graph is a function of the image gradient.

(5) Hierarchical clustering

Hierarchical clustering generic clustering algorithm family constructs nested clustering by successive merging or splitting them. The hierarchical structure of the cluster is represented as a tree (or tree graph). The root of a tree is the only cluster to collect all samples, while the leaf is the only cluster to collect only one sample.

Agglomerative clustering objects perform hierarchical clustering using a bottom-up approach: Each observation begins in its own clustering and the clustering is merged in turn. Linkage criteria determines metrics for merger strategies:

1. Ward minimizes the sum of squares of differences within all clusters. It is a variance minimization method. In this sense, it is similar to the K-means objective function, but it is processed by the cohesive hierarchy method.
2. Maximum or complete linkage minimizes the maximum distance between clusters and observations.
3. Average linkage minimizes the average distance between all observed values of a cluster pair.

Agglomerative clustering supports Ward, an average and complete connection strategy. Cohesive cluster has the behavior of "getting rich or getting rich," which leads to the uneven size of cluster. In this regard, complete contact is the worst strategy, and Ward gives the most conventional size. However, for Ward, affinity (or the distance used in clustering) cannot be changed, so for non-Euclidean metric, average join is a good choice.

(6) Density-based spatial clustering of applications with noise (DBSCAN)

DBSCAN is a typical density clustering algorithm. Compared with K-means and BIRCH, DBSCAN can be applied to both convex and non-convex samples.

This kind of density clustering algorithm generally assumes that the category can be determined by the compactness of the sample distribution. Samples of the same category are closely related to each other, that is to say, there must be samples of the same category not far from any sample of the same category. By grouping closely connected samples into one group, a clustering category is obtained. By dividing all the closely connected samples into different categories, we can get all the final clustering results.

DBSCAN is based on a set of neighborhoods to describe the compactness of the sample set. Parameters (ϵ, MinPts) are used to describe the compactness of the sample distribution in the neighborhood. Among them, ϵ describes the neighborhood distance threshold of a sample, and MinPts describes the threshold of the number of samples in the neighborhood where the distance of a sample is ϵ.

Assuming that my sample set is $D = (x_1, x_2, \ldots, x_m)$, the specific density description of DBSCAN is defined as follows:

1. ϵ-Neighborhood: For $x_j \in D$, its ϵ-Neighborhood contains a subset of sample set D whose distance from x_j is not greater than ϵ, i.e., $N_\varepsilon(x_j) = \{x_i \in D|$ distance $(x_i, x_j) \leq \epsilon\}$. The number of subsets is denoted as $|N_\varepsilon(x_j)|$.
2. Core object: For any sample $x_j \in D$, if its ϵ-neighborhood corresponding contains at least MinPts samples, that is, if $|N_\varepsilon(x_j)| \geq$ MinPts, then x_j is the core object.
3. Density direct: If x_i is located in the ϵ-neighborhood of x_j and x_j is the core object, it is said that x_i is from x_j density direct. Note that the converse is not necessarily true, that is, x_j cannot be said to be directly from x_i density at this time, unless and x_i is also the core object.
4. Density reachable: for x_i and x_j, if there are sample sequences $p_1, p_2, \ldots, p_T$ satisfying $p_1 = x_i$, $p_T = x_j$, and p_{T+} is directly from p_T density, x_j is said to be reachable from x_i density. That is to say, the density can satisfy the transmissibility. At this time, the transfer samples $p_1, p_2, \ldots, p_{T-1}$ in the sequence are all core objects, because only core objects can make other sample density direct. Notice that density reachability does not satisfy symmetry, which can be derived from the asymmetry of density direct access.
5. Density connection: For x_i and x_j, if there is core object sample x_k, so that both x_i and x_j can be reached by x_k density, then x_i and x_j density are connected. Attention: density connection is symmetrical.

It is easy to see from the following figure that the above definition is understood. MinPts $= 5$ and the red dots are the core objects, because there are at least five samples in the euro-neighborhood. The black sample is a non-core object. All samples with direct density of core objects are in hyperspheres with red core objects as the center. If they are not in hyperspheres, they cannot be directly dense. The core objects connected by green arrows in the figure constitute a sequence of samples with reachable density. In the $\epsilon\epsilon$-neighborhood of these density-reachable sample sequences, all samples are densely connected to each other.

Then we give the definition of DBSCAN: the maximum density connected sample set derived from the density reachability relation, which is a category or a cluster of our final clustering.

Next, we summarize the process of DBSCAN clustering algorithm.

Input: Sample set $D = (x_1, x_2, \ldots, x_m)$, neighborhood parameters (ϵ, MinPts), sample distance measurements.

Output: Cluster partitioning C.

1. Initialize the core object $\Omega = \emptyset$, initialize the cluster number $k = 0$, initialize the unvisited sample set $\Gamma = D$, and partition $C = \emptyset$.
2. For $j = 1, 2, \ldots, m$, follow the following steps to find out all the core objects:

    (a) Find the ϵ-neighborhood sub-sample set $N\epsilon(x_j)$ of sample x_j by distance measure.
    (b) If the number of samples in the sub-sample set satisfies $|N_\varepsilon(x_j)| \geq$ MinPts, the sample x_j is added to the core object sample set $\Omega = \Omega \cup \{x_j\}$.

3. If the set of core objects is $\Omega = \emptyset$, the algorithm ends, otherwise it goes to step 4).

4. In the core object set Ω, a core object o is randomly selected, the current cluster core object $\Omega_{cur} = \{o\}$ is initialized, the class number $k = k+1$ is initialized, the current cluster sample set $C_k = \{o\}$ is initialized, and the unreached sample set $\Gamma = \Gamma - \{o\}$ is updated.

5. If the current cluster core object $\Omega_{cur} = \emptyset$, then the current cluster C_k generation is completed, update the cluster partition $C = \{C_1, C_2, ..., C_k\}$, update the core object set $\Omega = \Omega - C_k$, go to step 3).

6. In the current cluster core object Ω_{cur}, a core object o' is extracted, and all ϵ-neighborhood sub-sample sets $N\epsilon(o')$ are found through the neighborhood distance thresholdϵ, so that $\Delta = N\epsilon(o') \cap \Gamma$, the current cluster sample set $C_k = C_k \cup \Delta$, the untried sample set $\Gamma = \Gamma - \Delta$, the updated $\Omega_{cur} = \Omega_{cur} \cup (N\epsilon(o') \cap \Omega)$ are transferred to step 5).

The output is: Cluster partition $C = \{C_1, C_2, ..., C_k\}$.

(7) Balanced iterative reducing and clustering using hierarchies (BIRCH)

BIRCH algorithm uses a tree structure to help us cluster quickly, which is similar to the balanced B^+ tree. It is generally called the clustering feature tree (CF tree). Each node of the tree is composed of several clustering features (CF). CF tree has several CFs for each node including leaf nodes, while CF for internal nodes has pointers to child nodes, and all leaf nodes are linked by a two-way list.

In the clustering feature tree, a clustering feature CF is defined as a triple, which can be represented by (N, LS, SS). Where N represents the number of sample points in this CF, which is understandable; LS represents the sum and vector of each feature dimension of the sample points in this CF, SS represents the square of each feature dimension of the sample points in this CF.

For CF tree, we generally have several important parameters. The first parameter is the maximum CF number B of each internal node, the second parameter is the maximum CF number L of each leaf node, and the third parameter is for a sample point in a CF of the leaf node. It is the maximum sample radius threshold R of each CF of the leaf node. That is to say, all sample points in this CF must be in the CF. A hypersphere with radius less than R.

Next, we describe how to generate CF tree. First, we define the parameters of CF tree: the maximum CF number B of internal nodes, the maximum CF number L of leaf nodes, and the maximum sample radius threshold R of each CF of leaf nodes. At the beginning, CF tree is empty and there are no samples. When a new sample point needs to be added to CF tree, the process of generating CF tree can be described as follows:

1. Searching for the nearest CF node in the leaf node and the leaf node closest to the new sample from the root node;

2. If the radius of the hypersphere corresponding to the CF node is still less than the threshold R after the new sample is added, all CF triples on the path are updated and the insertion ends. Otherwise, transfer to 3;

3. If the number of CF nodes of the current leaf node is less than the threshold L, a new CF node is created, a new sample is put, a new CF node is put into the leaf

node, all CF triples on the path are updated, and the insertion ends. Otherwise, transfer to 4;

4. The current leaf nodes are divided into two new leaf nodes, and the two CF tuples with the farthest distance from the hypersphere in all CF tuples of the old leaf nodes are selected as the first CF node of the two new leaf nodes. The other tuples and the new sample tuples are put into the corresponding leaf nodes according to the distance principle. Check the parent node to split up in turn, if it is necessary to split the leaf nodes in the same way as the leaf nodes.

This is the complete CF tree generation process. In fact, CF tree is established for all training set samples, and a basic BIRCH algorithm is completed. The corresponding output is a number of CF nodes, and the sample points in each node are a cluster. That is to say, the main process of BIRCH algorithm is the process of establishing CF tree. Now let us look at the process description of the BIRCH algorithm:

1. Read all the samples in turn and build a CF tree in memory. The method is referred to in the previous section.
2. (Optional)Select the CF tree established in the first step to remove some abnormal CF nodes, which usually have few sample points. Merge some hyperspheres with very close distances.
3. (Optional)Clustering all CF tuples using other clustering algorithms such as K-means to get a better CF tree. The main purpose of this step is to eliminate the unreasonable tree structure caused by the sample reading order and some tree structure splitting caused by the restriction of the number of CF nodes.
4. (Optional)The centroid of all CF nodes of CF tree generated in the third step is used as the initial centroid, and all sample points are clustered according to distance. This further reduces the unreasonable clustering caused by some constraints of CF tree.

As can be seen from the above, the key of BIRCH algorithm is step 1, which is the generation of CF tree. The other steps are to optimize the final clustering results.

(8)   Gaussian mixture model (GMM)

GMM, Gauss mixture model, is a linear superposition of several Gauss models. Each Gauss model is called component. GMM algorithm describes a distribution of data itself. If there are enough components, GMM can approximate any probability density distribution.

GMM is a linear superposition of K Gaussian distributions. The probability density function of GMM is as follows,

$$P(x) = \sum_{k=1}^{K} P(k)P(x|k) = \sum_{k=1}^{K} \pi_k N(x|\mu_k, \Sigma_k).$$

The above formula can be understood as: given a data set, we use GMM model to describe the distribution of the data set. When the data set is described by component

k, the probability density function of each data set is $P(x|k)$. There are K components in total. Each component contributes $P(k)$ to the data set generation, and its probability density function is $P(k)P(x|k)$. The probability density function of GMM is obtained by adding all components together.

If we want to randomly select a point from the GMM distribution, we can actually divide it into two steps: first, randomly select one of the K Gaussian components. The probability of each component being selected is actually its coefficient $P_i(k)$. After selecting the component, we can consider selecting a point from the distribution of the component separately. We have returned to the normal Gaussian distribution and transformed it into a known problem.

When GMM is used in cluster analysis, K components of GMM correspond to K clusters. Each component is a Gaussian distribution, its mean value is set to μ_k, variance is set to Σ_k, and the influence factor of this component is set to π_k. Then the GMM algorithm flow can be described as follows:

1. Estimate the probability that data is generated by each component (not the probability that each component is selected): for each data i, the probability that it is generated by component K is

$$g(i, k) = \frac{\pi_k N(x_i|\mu_k, \Sigma_k)}{\sum_{j=1}^{K} \pi_j N(x_i|\mu_j, \Sigma_j)},$$

where $N(x_i|\mu_k, \Sigma_k)$ is the posterior probability,

$$N(x|m, S) = \frac{1}{(2\pi)^{D/2}} \frac{1}{|\Sigma|^{1/2}} \exp\left\{-\frac{1}{2}(x - \mu)^T \Sigma^{-1}(x - \mu)\right\}.$$

2. By maximum likelihood estimation, the values of parameters μ_k and Σ_k can be obtained by calculating the derivative parameter $= 0$,

$$k_k = \frac{1}{N_k} \sum_{i=1}^{N} \gamma(i, k)x_i$$

$$\Sigma = \frac{1}{N_k} \sum_{i=1}^{N} \gamma(i, k)(x_i - \mu_k)(x_i - \mu_k)^T,$$

where $N_k = \sum_{i=1}^{N} \gamma(i, k)$, so π_k can be estimated as N_k/N.

3. Repeat the first two steps until the value of the likelihood function converges.

References

1. Pedregosa F, Varoquaux G, Gramfort A, Michel V, Thirion B, Grisel O, Blondel M, Prettenhofer P, Weiss R, Dubourg V, Vanderplas J, Passos A, Cournapeau D, Brucher M, Perrot M, Duchesnay E (2011) Scikit-learn: machine learning in Python. J Mach Learn Res 12:2825–2830
2. http://scikit-learn.org/stable/auto_examples/tree/plot_tree_regression.html#sphx-glr-auto-exa mples-tree-plot-tree-regression-py
3. http://scikit-learn.org/stable/modules/tree.html
4. http://scikit-learn.org/stable/auto_examples/plot_multioutput_face_completion.html#sphx-glr-auto-examples-plot-multioutput-face-completion-py
5. Yang X, Zhou P, Wang M (2019) Person reidentification via structural deep metric learning. IEEE Trans Neural Netw Learn Syst 30(10):2987–2998
6. http://scikit-learn.org/stable/modules/gaussian_process.html#gaussian-process-regression-gpr
7. http://scikit-learn.org/stable/auto_examples/svm/plot_iris.html#sphx-glr-auto-examples-svm-plot-iris-py

Chapter 5
State-of-the-Art Artificial Intelligence Algorithms

5.1 Deep Learning

Deep learning is a modern update to ANN that exploits abundant cheap computation. Usually, they are concerned with building much larger and more complex neural networks and many methods are concerned with semi-supervised learning problems where large data sets contain very little labeled data.

It is a part of a broader family of machine learning methods based on learning data representations, as opposed to task-specific algorithms. Some representations are loosely based on interpretation of information processing and communication patterns in a biological nervous system, such as neural coding that attempts to define a relationship between various stimuli and associated neuronal responses in the brain.

Deep learning architectures such as deep neural network, deep belief network, and recurrent neural network have been applied to fields including computer vision, speech recognition, natural language processing, audio recognition, social network filtering, machine translation, bioinformatics and drug design, where they have produced results comparable to and in some cases, superior to human experts.

Neural networks can be classified from different dimensions. According to data flow, it can be divided into feed-forward neural network, recursive neural network, back-forward neural network. According to the form of neuron organization in the network, it can be divided into full-connected network and partial-connected network. According to the behavior and connection of neurons in the network, it can be divided into simple (fully connected) neural network, convolutional neural network, cyclic neural network, and recurrent neural network. According to the training method, it can be divided into supervised learning, unsupervised learning, and reinforcement learning.

Different neural networks have different applications. Fully connected networks are often used in data analysis and are an important part of other networks. Convolutional neural networks are often used in computer vision and data with local correlation. Cyclic neural networks are often used in natural language processing and data with sequence and correlation.

© Tsinghua University Press 2020
X. Liang, *Social Computing with Artificial Intelligence*,
https://doi.org/10.1007/978-981-15-7760-4_5

This book will briefly introduce the three most widely used models: deep neural network, convolutional neural network, and recurrent neural Network according to the different behaviors and connections of neurons in the network.

5.1.1 Deep Neural Network

Deep neural network is the foundation of deep learning. This book will describe deep neural network step by step from the following steps.

(1) From perceptron to neural network

Perceptron is a model with several inputs and one output. It is a basic unit in deep neural network.

A linear relationship is learned between output and input, and intermediate output results are obtained,

$$z = \sum_{i=1}^{m} w_i x_i + b.$$

Then, a neuron activation function is connected. Generally speaking, the activation function used by the perceptron is unit function,

$$f(x) = \text{unit}(x) = \begin{cases} -1, & x < 0, \\ 1, & x \geq 0. \end{cases}$$

So, we can get the output result 1 or -1 that we want. This model can only be used for binary classification, and cannot learn more complex nonlinear models, so it cannot be used in industry.

Neural network is extended on the model of perceptron. There are three main points in summary:

1. Hidden layer is added. Hidden layer can have many layers, which can enhance the expressive ability of the model. The following examples show that the complexity of so many hidden layer models has increased a lot (see Fig. 5.1).
2. Neurons in the output layer can have more than one output and multiple outputs, so that the model can be flexibly applied to classification and regression, and other machine learning fields such as dimensionality reduction and clustering. An example of the output layer of multiple neurons is shown below. There are now four neurons in the output layer (see Fig. 5.2).
3. To extend the activation function, the activation function of perceptron is unit(x), which is simple but has limited processing power. Therefore, the activation function commonly used in neural networks, such as sigmoid function used in logistic regression, is

Fig. 5.1 Feed-forward
neural network

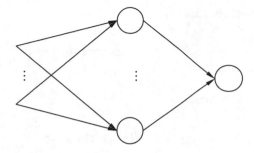

Fig. 5.2 Neural networks
with multiple neuron outputs

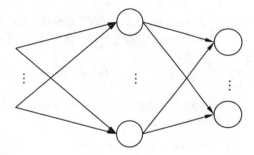

$$f(x) = \text{sigmoid}(x) = \frac{1}{1 + \exp^{-x}}.$$

In addition, other commonly used activation functions include

tanh function

$$f(x) = \tanh(x) = \frac{\exp^{x} - \exp^{-x}}{\exp^{x} + \exp^{-x}},$$

relu function

$$f(x) = \text{relu}(x) = \begin{cases} x, & x > 0, \\ 0, & x \leq 0, \end{cases}$$

prelu function

$$f(x) = \text{prelu}(x) = \begin{cases} x, & x > 0, \\ ax, & x \leq 0, \end{cases}$$

maxout function

$$f(x) = \max(w_1^{\text{T}} x + b_1, w_2^{\text{T}} x + b_2, \ldots, w_n^{\text{T}} x + b_n),$$

elu function

$$f(x) = \mathrm{elu}(x) = \begin{cases} x, & x > 0, \\ a(\exp^x - 1), & x \leq 0, \end{cases}$$

selu function

$$f(x) = \mathrm{selu}_{\mathrm{scale}} = \begin{cases} x, & x > 0, \\ \mathrm{selu}_a(\exp^x - 1), & x \leq 0, \end{cases}$$

with $\begin{cases} \mathrm{selu}_{\mathrm{scale}} = 1.0507009873554804934193349852946 \\ \mathrm{selu}_a = 1.6732632423543772848170429916717 \end{cases}$, and swish function

$$f(x) = \mathrm{swish}(x) = x \cdot \mathrm{sigmoid}(x).$$

The use of different activation functions further enhances the expressive ability of neural networks.

(2) **Deep neural network (DNN)**

Deep neural network can be understood as a neural network with multiple hidden layers. It should be pointed out that another well-known model is called multilayer perceptron, which has the same basic structure as deep neural network. The only difference is that the activation function of multilayer perceptron is unit function, while the activation function of deep neural network is sigmoid, tanh, and other nonlinear functions.

The neural network layer in DNN can be divided into three types: input layer, hidden layer, and output layer. As shown in Fig. 5.3, the first layer is input layer, the last layer is output layer, and the number of layers in the middle is hidden layer (see Fig. 5.3).

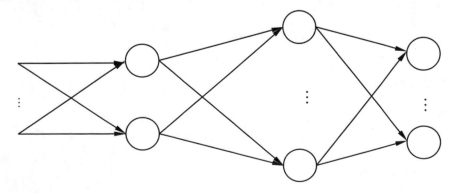

Fig. 5.3 Deep neural network

The layers are fully connected, that is, any one of the ith layers must be connected to any one of the $(i + 1)$th layers. Although DNN seems very complex, it is the same as the perceptron in small local models, that is, after weighted summation of input elements, it is input into an activation function to get an output result.

In order to describe the deep neural network more accurately, a simple three-layer neural network in Fig. 5.2 is taken as an example to define parameters. The input layer of the network is n-dimensional, $w_{ij}^{(l)}$ indicating the weight of the ith neuron of the first layer connected to the jth input. $W^{(l)} \in R[p, k]$ is a weight matrix. There are p neurons in the lth layer. The weight of each neuron is a k-dimensional vector, and k is the number of $(l - 1)$th layer neurons. The output is

$$y^{(l)} = \text{output}^{(l)} = f(W \cdot \text{output}^{(l-1)} + b).$$

(3) **DNN feed-forward propagation**

We take a simple three-tier neural network as an example to give readers a most intuitive understanding of the forward propagation process of deep neural network. A three-layer neural network is defined. The input is three variables: x_1, x_2, x_3, the hidden layer is three neurons, and the weight coefficient is w, the sum of the weights before the input of the neuron activation function is z, the activation function is $\sigma(z)$, and the output value is $a = \sigma(z)$. The output layer has an output unit, and the output value is also expressed by a.

For the output of the second layer a_1^2, a_2^2, a_3^2, we have

$$a_1^2 = \sigma(z_1^2) = \sigma(w_{11}^2 x_1 + w_{12}^2 x_2 + w_{13}^2 x_3 + b_1^2),$$

$$a_2^2 = \sigma(z_2^2) = \sigma(w_{21}^2 x_1 + w_{22}^2 x_2 + w_{23}^2 x_3 + b_2^2),$$

$$a_3^2 = \sigma(z_3^2) = \sigma(w_{31}^2 x_1 + w_{32}^2 x_2 + w_{33}^2 x_3 + b_3^2).$$

For the third-level output a_1^3, we have

$$a_1^3 = \sigma(z_1^3) = \sigma(w_{11}^3 a_1^2 + w_{12}^3 a_2^2 + w_{13}^3 a_3^2 + b_1^3).$$

Generalizing the above example, supposing that there are m neurons in the $(l - 1)$th layer, we have the following for the output of the jth neuron in the lth layer,

$$a_j^l = \sigma(z_j^l) = \sigma\left(\sum_{k=1}^{m} w_{jk}^l a_k^{l-1} + b_j^l\right).$$

As can be seen from the above, the use of algebraic methods to represent output one by one is more complex, and if the use of matrix law can be succinct. Assuming that there are m neurons in the $(l - 1)$th layer and n neurons in the lth layer, the linear coefficients of the lth layer w form an $n \times m$ matrix W^l, the bias of the lth layer b forms an $n \times 1$ vector b^l, the output of the $(l-1)$th layer a forms an $m \times 1$ vector a^{l-1}, the inactive prelinear output of the lth layer forms a vector z^l, and the output of the lth layer a forms an $n \times 1$ vector a^l. The output of the lth layer is as follows:

$$a^l = \sigma(z^l) = \sigma(W^l a^{l-1} + b^l).$$

This representation is simple and beautiful, and we will use matrix method to express it later.

In summary, the forward propagation algorithm of DNN uses several weight coefficient matrices W, bias vectors b, and input value vectors x to perform a series of linear operations and activation operations, starting from the input layer, backward calculation from one layer to the output layer, until the output results are obtained.

Input: The total number of layers L, the matrix W corresponding to all hidden layers and output layers, bias vectors b, input value vectors x.

Output: Output of the output layer a^L.

1. Initialize $a^1 = x$.
2. For $l = 2$ to L, calculate

$$a^l = \sigma(z^l) = \sigma(W^l a^{l-1} + b^l).$$

The final result is output a^L.

(4) **DNN back-propagation**

Before DNN backpropagation algorithm, we need to select a loss function to measure the loss between the calculated output of training samples and the real output of training samples. There are many choices of loss function. The commonly used loss function is the mean square deviation function. For each sample, the loss function can be expressed as follows:

$$J(W, b, x, y) = \frac{1}{2} \|a^L - y\|_2^2,$$

where y is the vector of the feature output, and $\|S\|_2$ is the L_2 regulation of S.

Next, we use gradient descent method to iteratively update each layer W and b.

The first is the L layer of the output layer W, b, noting that the output layer W, b satisfies the following formula,

$$a^L = \sigma(z^L) = \sigma(W^L a^{L-1} + b^L).$$

Thus, for the parameters of the output layer, our loss function becomes,

$$J(W, b, x, y) = \frac{1}{2} \|a^L - y\|_2^2 = \frac{1}{2} \|\sigma(W^L a^{L-1} + b^L) - y\|_2^2.$$

In this way, the gradient solution of W, b becomes simple,

$$\frac{\partial J(W, b, x, y)}{\partial W^L} = \frac{\partial J(W, b, x, y)}{\partial z^L} \frac{\partial z^L}{\partial W^L} = (a^L - y) * \sigma'(z^L)(a^{L-1})^T,$$

$$\frac{\partial J(W, b, x, y)}{\partial b^L} = \frac{\partial J(W, b, x, y)}{\partial z^L} \frac{\partial z^L}{\partial b^L} = (a^L - y) * \sigma'(z^L).$$

Notice that there is a sign *, which represents the Hadamard product and can also be recorded as $\odot$. For two vectors of the same dimension $A(a_1, a_2, \ldots, a_n)^T$ and $B(b_1, b_2, \ldots, b_n)^T$, then $A*B = (a_1 b_1, a_2 b_2, \ldots, a_n b_n)^T$.

We noticed that there is a common part $\frac{\partial J(W, b, x, y)}{\partial z^L}$ in solving the output layer W, b, so we can calculate the common part, that is, calculate z^L first, which is calculated as,

$$\delta^L = \frac{\partial J(W, b, x, y)}{\partial z^L} = (a^L - y) * \sigma'(z^L).$$

After calculating the gradient of the output layer, the gradient of the layer, the layer, etc., needs to be gradually deduced. For the inactive output of the first layer, its gradient can be expressed as:

After calculating the gradient of the output layer in this way, it is still necessary to gradually push out the gradient of the $(L - 1)$th layer, $(L - 2)$th layer. For the inactive output z^L of the l layer, its gradient can be expressed as,

$$\delta^l = \frac{\partial J(W, b, x, y)}{\partial z^L} = \frac{\partial J(W, b, x, y)}{\partial z^L} \frac{\partial z^L}{\partial z^{L-1}} \frac{\partial z^{L-1}}{\partial z^{L-2}} \cdots \frac{\partial z^{l+1}}{\partial z^l}.$$

If we can calculate the δ^l of the lth layer in turn, it is easy to calculate W^l, b^l of this layer, because $z^l = W^l a^{l-1} + b^l$, so we can easily calculate the gradient of the lth layer W^l, b^l as follows:

$$\frac{\partial J(W, b, x, y)}{\partial W^l} = \frac{\partial J(W, b, x, y)}{\partial z^l} \frac{\partial z^l}{\partial W^l} = \delta^l (a^{l-1})^T,$$

$$\frac{\partial J(W, b, x, y)}{\partial b^l} = \frac{\partial J(W, b, x, y)}{\partial z^l} \frac{\partial z^l}{\partial b^l} = \delta^l.$$

It can be seen that the solution δ^l is the key to solve the gradient W^l, b^l. Because we get,

$$\delta^l = \frac{\partial J(W, b, x, y)}{\partial z^l} = \frac{\partial J(W, b, x, y)}{\partial z^{l+1}} \frac{\partial z^{l+1}}{\partial z^l} = \delta^{l+1} \frac{\partial z^{l+1}}{\partial z^l}.$$

So, the key to using the inductive method to deduce δ^{l+1} and δ^l is to solve $\frac{\partial z^{l+1}}{\partial z^l}$.

From $z^{l+1} = W^{l+1}a^l + b^{l+1} = W^{l+1}\sigma(z^l) + b^{l+1}$, we can find that,

$$\frac{\partial z^{l+1}}{\partial z^l} = (W^{l+1})^{\mathrm{T}} * \underbrace{(\sigma'(z^l), \ldots, \sigma'(z^l))}_{nl+1}.$$

By introducing the formulas above into the formulas δ^{l+1} and δ^l, it can be concluded that,

$$\delta^l = \delta^{l+1}\frac{\partial z^{l+1}}{\partial z^l} = (W^{l+1})^{\mathrm{T}}\delta^{l+1} * \sigma'(z^l).$$

Now, we get the recursive formula of δ^l. As long as the δ^l of a certain layer is found, solving the corresponding gradient of W^l, b^l is very simple.

Now, we summarize the process of DNN backpropagation algorithm. Since the gradient descent method has three variants: batch, mini-Batch and random, in order to simplify the description, we take the most basic batch gradient descent method as an example to describe the back propagation algorithm. In fact, the mini-Batch gradient descent method is the most widely used method in the industry. But the difference lies only in the selection of training samples during iteration.

Input: Total number of layers L, and number of neurons in each hidden layer and output layer, activation function, loss function, iteration step size α, maximum iteration number max and stop iteration threshold ϵ, input m training samples $\{(x_1, y_1), (x_2, y_2), \ldots, (x_m, y_m)\}$.

Output: Linear relationship coefficient matrix W and bias vector b for each hidden layer and output layer.

1.   Initialize the values of the linear relationship coefficient matrix W and bias vector b of each hidden layer and the output layer to be a random value.
2.   for iter to 1 to max:
2-1. for $i = 1$ to m:
(A)   Set DNN input a^l to x_i.
(B)   For $l = 2$ to L, perform forward propagation algorithm to calculate

$$a^{i,l} = \sigma(z^{i,l}) = \sigma(W^l a^{i,l-1} + b^l).$$

(C)   Calculating $\delta^{i,L}$ of output layer by loss function.
(D)   For $l = 2$ to 2, perform forward propagation algorithm to calculate

$$\delta^{i,L} = (W^{l+1})^{\mathrm{T}}\delta^{i,l+1} * \sigma'(z^{i,l}).$$

2-2. for $l = 2$ to L, update W^l, b^l of the lth layer

$$W^l = W^l - \alpha \sum_{i=1}^{m} \delta^{i,l}(a^{i,l-1})^{\mathrm{T}},$$

$$b^l = b^l - \alpha \sum_{i=1}^{m} \delta^{i,l}.$$

2-3. If the change value of all W, b is less than the stop iteration threshold, jump out of the iteration cycle to step 3.

3. The linear relationship coefficient matrix W and bias vector b of each hidden layer and output layer are output.

5.1.2 Convolutional Neural Network

(1) Convolutional neural network structure

Convolutional neural networks have many classical models, including LeNet5 (1998), Alexnet (2012), Google Net (2014), VGG (2014), and deep residual learning (2015). We take LeNet model proposed by Lecun in 1998 as an example to introduce the basic knowledge of CNN [1]. A typical LeNet structure is shown in the following figure. It consists of two convolution layers, two pooling layers and two fully connected layers. Convolutions are 5 * 5 templates, stride $= 1$, and pooling is max. Next, we introduce the process of convolution layer, pooling layer and full connection layer.

(2) Convolution layer

For image convolution, it is actually to multiply the elements of different local matrices of the output image and the positions of the convolution kernel matrix, and then add them together. For example, the input in the graph is a two-dimensional matrix of 3×4, and the convolution kernel is a matrix of 2×2. Here, we assume that the convolution is convoluted by moving one pixel at a time. First, we multiply the 2×2 local and convolution kernel convolutions in the upper left corner of the input, that is, the elements of each position are multiplied and then added, and the obtained element S_{00} of the output matrix S has the value $aw + bx + ey + fz$. Then, we translate the input part to the right by one pixel. Now, we convolute the matrix and the convolution kernel is composed of four elements (b, c, f, g). So, we get the elements S_{01} of the output matrix S. In the same way, we can get the elements S_{02}, S_{10}, S_{11}, and S_{12} of the output matrix S. Finally, we get that the convolution output matrix is a 2×3 matrix S.

The examples given above are all two-dimensional inputs. The convolution process is relatively simple. What if the input is multidimensional? For example, in the previous set of convolution layer + pooling layer, the output is three matrices, and these three matrices are input. So how do we deconvolute?

For example, a typical color image is an input matrix corresponding to R, G, and B dimensions, i.e., the input is an $x \times y \times z$ three-dimensional tensor. The final convolution process is similar to that of the two-dimensional matrix above. The former is the convolution of the matrix, that is, the elements of the corresponding positions of the two matrices are multiplied and added. The latter is the convolution of tensors, that is, after the convolution of three sub-matrices of two tensors, the results of convolution are added together and the bias b is added.

First, we need to define several nouns in CNN:

- Convolution kernel: A tensor that convolutes an image.
- Convolved feature: A new matrix obtained by the convolution of image and convolution kernel, i.e., output volume in the graph. There are several convolution kernel and several convolved features.
- Receptive field: The location covered by convolution kernel.
- Stride: The number of units that convolution kernel slides each time in image. For example, stride = [2, 2] means convolution kernel moves two units to the right and two units to the downward.
- Padding: When padding = valid denotes an edge to it, that is, the edge of receptive field aligns with the edge of image to begin convolution. When padding = sample denotes that it is convoluted by the center, that is, the positive center of receptive field and the edge of image, the rest is filled with 0. The image of the original figure $5 \times 5 \times 3$ is complemented to the tensor of $7 \times 7 \times 3$.

Taking an input image of $3 \times 5 \times 5$ as an example, using two $3 \times 3 \times 3$ convolution kernels, the convolution step is 2, and the output is a tensor of $3 \times 3 \times 2$, as shown in the following figure. If the above convolution process is expressed by mathematical formulas, it is as follows:

$$s(i, j) = (X * W)(i, j) + b = \sum_{k=1}^{n\_in} (X_k * W_k)(i, j) + b,$$

where n_in is the number of input matrices, or the dimension of the last dimension of the tensor. X_k represents the kth input matrix, and W_k represents the kth sub-convolution kernel matrix of the convolution kernel. $s(i, j)$, that is, the value of the corresponding position element of the output matrix corresponding to the convolution kernel W. For convoluted output, the relu activation function is used to change the element value corresponding to the position less than 0 in the output tensor to zero.

(3) **Pooling**

Compared with the convolution layer, the pooling layer is much simpler. Pooling actually completes the process of compressing each sub-matrix of input tensor. For example, the pooling of an $N \times N$ reduces the dimensionality of the input matrix by compressing each $n \times n$ element of the sub-matrix into one element. There are two common pooling criteria, max pooling and average pooling, which take the maximum or average value of the corresponding region as the element value after pooling.

(4) **CNN feed-forward propagation**

In the previous article, we have already talked about the structure of CNN, including the output layer, several convolutional layers + relu activation functions, several pooling layers, DNN fully connected layers, and finally, the output layer using the softmax activation function. Here, we use a color car sample to identify the structure of the CNN from a sensory review. The CONV in the figure is the convolution layer, POOL is the pooling layer, and FC is the DNN fully connected layer, including the last output layer of the softmax activation function. To straighten out CNN's forward propagation algorithm, the focus is on the forward propagation of the input layer, the forward propagation of the convolutional layer, and the forward propagation of the pooling layer. The forward propagation algorithm of the DNN fully connected layer and the output layer using the softmax activation function has already been mentioned in the DNN.

1. **Forward propagation of input layer**

Forward propagation of input layer is the first step of CNN forward propagation algorithm. Generally, the input layer corresponds to the convolution layer, so our title is that the input layer propagates forward to the convolution layer. Consider the simplest, two-dimensional black and white images as an example. In this way, the input layer X is a matrix whose value is equal to the value of each pixel position in the picture. At this time, the convolution kernel W connected with the convolution layer is also a matrix.

If the samples are all color images with RGB, then the input X is three matrices, that is, the matrices corresponding to R, G, and B, or a tensor. At this time, the convolution kernel connected with the convolution layer W is also a tensor, and the corresponding dimension of the last dimension is 3. That is, each convolution kernel is composed of three sub-matrices. In the same way, for samples such as 3D color images, our input can be four-dimensional and five-dimensional tensors, and the corresponding convolution kernel W is also a high-dimensional tensor.

Regardless of the dimension, for our input, the forward propagation process can be expressed as,

$$a^2 = \sigma(z^2) = \sigma(a^1 * W^2 + b^2).$$

Among them, the superscript represents the number of layers, * represents the convolution, and b represents our bias. σ is an activation function, and it is generally relu here.

2. Hidden layer propagates forward to convolution layer

Assuming that the output of the hidden layer is a three-dimensional tensor corresponding to M matrices, the convolution kernel output to the convolution layer is also a three-dimensional tensor corresponding to M sub-matrices. At this point, the expression is very similar to the input layer, and the same is true,

$$a^l = \sigma(z^l) = \sigma(a^{l-1} * W^l + b^l).$$

It can also be written in the form of additions of corresponding positions after convolution of M sub-matrices, i.e.,

$$a^l = \sigma(z^l) = \sigma\left(\sum_{k=1}^{M} z_k^l\right) = \sigma\left(\sum_{k=1}^{M} a_k^{l-1} * W_k^l + b^l\right).$$

The input here comes from the hidden layer, not the matrix formed by the original image sample we input. Here, we need to define the number of convolution kernels K, dimension F of convolution kernels matrix, fill size P, and step size S.

3. Hidden layer propagates forward to pooling layer

The processing logic of the pooling layer is relatively simple. Our purpose is to reduce and generalize the input matrix. For example, the input matrices are $N \times N$ dimensional, and the output matrices are $k \times k$-dimensional if our pooling size is an $\frac{N}{k} \times \frac{N}{k}$ region.

The CNN model parameters that we need to define here are:

(a) Size of pooling area K.
(b) The standard of pooling is usually max or average.

4. Hidden layer propagates forward to full connection layer

Since the full connection layer is a common DNN model structure, we can directly use the forward propagation algorithm logic of DNN, that is

$$a^l = \sigma(z^l) = \sigma(W^l a^{l-1} + b^l).$$

The activation function here is usually sigmoid or tanh. After several full connection layers, the last layer is the softmax output layer. The only difference between the output layer and the normal full connection layer is that the activation function is a softmax function.

The CNN model parameters that we need to define here are:

(a) Activation function of full connection layer.
(b) Number of neurons in all layers of the connective layer.

5. Summary of CNN forward propagation algorithms

With the above foundation, we now summarize the forward propagation algorithm of CNN.

Input: 1 Image sample, the number of layers L of CNN model, and the type of all hidden layers. For convolution layer, we need to define the size K of convolution kernel, dimension F of convolution kernel matrix, fill size Q, step S. For the pooling layer, the size of pooling area K and the pooling standard (max or average) should be defined. For the full connection layer, the activation function of the full connection layer (except the output layer) and the number of neurons in each layer should be defined.

Output: Output of CNN model a^L.

(a) According to the filling size Q of the input layer, the edge of the original image is filled and the input tensor a^1 is obtained.
(b) Initialize all hidden layer parameters W and b.
(c) For $l = 2$ to $L - 1$:

(i) If the lth layer is a convolution layer, the output is

$$a^l = \text{ReLU}(z^l) = \text{ReLU}(a^{l-1} * W^l + b^l).$$

(ii) If the lth layer is the pooling layer, then the output is $a^l = \text{pool}(a^{l-1})$, where the pool refers to the process of reducing the input tensor according to the size of the pooling area K and the pooling criteria.
(iii) If the lth layer is a fully connected layer, the output is

$$a^l = \sigma(z^l) = \sigma(W^l a^{l-1} + b^l).$$

(d) For the lth layer of the output layer,

$$a^L = \text{softmax}(z^L) = \text{softmax}(W^L a^{L-1} + b^L).$$

6. CNN back forward propagation

We can apply the back propagation algorithm of DNN to calculate the back propagation process of CNN. But obviously, there are some differences in CNN, so we cannot directly apply the formula of DNN backpropagation algorithm. Specifically, there are several issues that need to be addressed:

(a) There is no activation function in the pooling layer. This problem is better solved. We can make the activation function of the pooling layer $\sigma(z) = z$, that is, it is itself after activation. Thus, the derivative of the activation function of the pooling layer is 1.
(b) When the pooling layer propagates forward, it compresses the input, so we need to deduce δ^{l-1} forward and backward now. This deduction method is completely different from DNN.

(c) The convolution layer is the output of the current layer obtained by tensor convolution, or the sum of several matrix convolutions. This is very different from DNN. The full connection layer of DNN is to get the output of the current layer directly by matrix multiplication. In this way, when the convolution layer propagates backward, the calculation method of δ^{l-1} recursion in the upper layer must be different.

(d) For convolution layer, since w uses convolution, the way to derive W, b of all convolution kernels from δ^l is different.

As can be seen from the above, problem 1 is a better solution, but problems 2, 3 and 4 need to be well brainstormed, and problems 2, 3 and 4 are also the key to solve CNN backpropagation algorithm. In addition, we should note that a_1 and z_1 in DNN are only vectors, while a_l and z_l in CNN are tensors, which are three-dimensional, that is, composed of several input sub-matrices.

Next, we will study the back propagation algorithm of CNN step by step for problems 2, 3, and 4. In the process of research, it should be noted that because the convolution layer can have multiple convolution kernels, the processing methods of each convolution kernel are identical and independent. In order to simplify the complexity of the algorithm formula, we hereinafter mention that the convolution kernel is one of several convolution kernels in the convolution layer.

7. **Knowing the δ^l of the pooling layer, deriving the δ^{l-1} of the previous hidden layer**

We first solve the problem 2 above. If we know the pooling layer δ^l, we derive the upper hidden layer δ^{l-1}.

In the forward propagation algorithm, the pooling layer usually uses max or average to pool the input, and the size of the pooling area is known. Now, in turn, we need to restore the error corresponding to the previous larger region from the reduced error δ^l.

In reverse propagation, we first reduce the size of all the sub-matrices of δ^l to the size before pooling, and then, if max, we place the values of each pooling locality of all the sub-matrices of δ^l at the position where the maximum value is obtained by the forward propagation algorithm. In the case of average, the values of each pooling locality of all the sub-matrices of δ^l are averaged and then placed at the position of the sub-matrices after reduction. This process is generally called up sample.

An example can be used to illustrate this conveniently: let us assume that the size of our pooling region is 2×2. The kth sub-matrix of δ^l is

$$\delta_k^l = \begin{pmatrix} 2 & 8 \\ 4 & 6 \end{pmatrix}.$$

Since the pooled area is 2×2, we first restore δ_k^l, which becomes,

$$\begin{pmatrix} 0\,0\,0\,0 \\ 0\,2\,8\,0 \\ 0\,4\,6\,0 \\ 0\,0\,0\,0 \end{pmatrix}.$$

If it is max, suppose that the maximum positions we recorded in the forward propagation are upper left, lower right, upper right, and lower left, respectively, then the converted matrix is

$$\begin{pmatrix} 2\,0\,0\,0 \\ 0\,0\,0\,8 \\ 0\,4\,0\,0 \\ 0\,0\,6\,0 \end{pmatrix}.$$

In the case of average, the average is performed: the converted matrix is

$$\begin{pmatrix} 0.5\ 0.5\ 2\ \ \ 2 \\ 0.5\ 0.5\ 2\ \ \ 2 \\ 1\ \ \ 1\ \ \ 1.5\ 1.5 \\ 1\ \ \ 1\ \ \ 1.5\ 1.5 \end{pmatrix}.$$

So, we get the value of the upper layer $\frac{\partial J(W,b)}{\partial a_k^{l-1}}$, to get δ_k^{l-1},

$$\delta_k^{l-1} = \frac{\partial J(W,b)}{\partial a_k^{l-1}} \frac{\partial a_k^{l-1}}{\partial z_k^{l-1}} = \text{upsample}(\delta_k^l) * \sigma'(z_k^{l-1}).$$

Among them, upsample function completes the logic of pooling error matrix amplification and error redistribution.

In summary, for tensor δ^{l-1}, we have

$$\delta^{l-1} = \text{upsample}(\delta^l) * \sigma'(z^{l-1}).$$

8. **Knowing the δ^l of the convolutional layer, deriving the δ^{l-1} of the upper hidden layer**

For the backward propagation of convolution layer, we first recall the forward propagation formula of convolution layer,

$$a^l = \sigma(z^l) = \sigma(a^{l-1} * W^l + b^l).$$

In DNN, we know that the recursive relationship between δ^{l-1} and δ^l is

$$\delta^l = \frac{\partial J(W,b,x,y)}{\partial z^l} = \frac{\partial J(W,b,x,y)}{\partial z^{l+1}} \frac{\partial z^{l+1}}{\partial z^l} = \delta^{l+1} \frac{\partial z^{l+1}}{\partial z^l}.$$

Therefore, in order to derive the recursive relationship between δ^{l-1} and δ^l, the gradient expression of $\frac{\partial z^l}{\partial z^{l-1}}$ must be calculated.

It is noted that the relationship between z^{l-1} and z^l is as follows:

$$z^l = a^{l-1} * W^l + b^l = \sigma(z^{l-1}) * W^l + b^l.$$

So, we have

$$\delta^{l-1} = \delta^l \frac{\partial z^l}{\partial z^{l-1}} = \delta^l * \mathrm{rot}180(W^l) * \sigma'(z^{l-1}).$$

The formula here is actually similar to DNN. The difference is that the convolution kernel is rotated 180° when deriving the formula with convolution. That is, rot180() in formula, flip 180° means flip up and down once, then turn left and right once. In DNN, this is just a matrix transposition. So why? Because there are tensors, there are too many parameters to deduce directly. Let us take a simple example to illustrate why the convolution kernel needs to be flipped after derivation.

Assuming that the output a^{l-1} of layer $l-1$ is a 3×3 matrix and the convolution kernel W^l of layer l is a 2×2 matrix, the output z^l is a 2×2 matrix with a 1-pixel step. If we simplify b^l to zero, then we have

$$a^{l-1} * W^l = z^l.$$

We list the matrix expressions of a, W, z as follows:

$$\begin{pmatrix} a_{11} & a_{12} & a_{13} \\ a_{21} & a_{22} & a_{23} \\ a_{31} & a_{32} & a_{33} \end{pmatrix} * \begin{pmatrix} w_{11} & w_{12} \\ w_{21} & w_{22} \end{pmatrix} = \begin{pmatrix} z_{11} & z_{12} \\ z_{21} & z_{22} \end{pmatrix}.$$

Using the definition of convolution, it is easy to conclude that,

$$z_{11} = a_{11}w_{11} + a_{12}w_{12} + a_{21}w_{21} + a_{22}w_{22},$$

$$z_{12} = a_{12}w_{11} + a_{13}w_{12} + a_{22}w_{21} + a_{23}w_{22},$$

$$z_{21} = a_{21}w_{11} + a_{22}w_{12} + a_{31}w_{21} + a_{32}w_{22},$$

$$z_{22} = a_{22}w_{11} + a_{23}w_{12} + a_{32}w_{21} + a_{33}w_{22}.$$

Then, we simulate reverse derivation,

$$\nabla a^{l-1} = \frac{\partial J(W, b)}{\partial a^{l-1}} = \frac{\partial J(W, b)}{\partial z^l} \frac{\partial z^l}{\partial a^{l-1}} = \delta^l \frac{\partial z^l}{\partial a^{l-1}}.$$

As can be seen from the above formula, for a^{l-1}, the gradient error ∇a^{l-1} is equal to the gradient error of layer l multiplied by $\frac{\partial z^l}{\partial a^{l-1}}$ and $\frac{\partial z^l}{\partial a^{l-1}}$ corresponds to the value of w associated with the above example. Assuming that the backpropagation error of our z matrix is a 2×2 matrix composed of δ_{11}, δ_{12}, δ_{21}, and δ_{22}, we can write out the gradients of nine scalars of ∇a^{l-1} by using the formulas and four equality of the above gradients.

For example, for the gradient of a_{11}, since a_{11} is only a product of z_{11} in four equations, we have

$$\nabla a_{11} = \delta_{11} w_{11}.$$

For example, for the gradient of a_{12}, because a_{11} and z_{12}, z_{11} have a product relation in four equations, we have

$$\nabla a_{12} = \delta_{11} w_{12} + \delta_{12} w_{11}.$$

In the same way, we get the rest. In the form of a matrix convolution, it can be expressed as,

$$\begin{pmatrix} 0 & 0 & 0 & 0 \\ 0 & \delta_{11} & \delta_{12} & 0 \\ 0 & \delta_{21} & \delta_{22} & 0 \\ 0 & 0 & 0 & 0 \end{pmatrix} * \begin{pmatrix} w_{22} & w_{21} \\ w_{12} & w_{11} \end{pmatrix} = \begin{pmatrix} \nabla a_{11} & \nabla a_{12} & \nabla a_{13} \\ \nabla a_{21} & \nabla a_{22} & \nabla a_{23} \\ \nabla a_{31} & \nabla a_{32} & \nabla a_{33} \end{pmatrix}.$$

In order to fit the gradient calculation, we fill a circle of 0 around the error matrix. At this time, we convolute the gradient errors of the convolution kernel after flipping and reverse propagation, and get the previous gradient errors. This example intuitively introduces the reason why the convolution kernel needs to flip $180°$ when deriving a formula containing convolution.

9. **Knowing δ^l of the convolutional layer, and deriving the gradient of W, bW and b of the layer**

We can now deduce the gradient error δ^l of each layer. For the fully connected layer, we can calculate the gradient of layer W, b according to the back propagation algorithm of DNN, and the gradient of layer W, b does not exist in the pooling layer, nor does it need to be calculated in the pooling layer. Only W, b of the convolution layer needs to be found.

It is noted that the relationship between convolution z and W, b is as follows:

$$z^l = a^{l-1} * W^l + b.$$

So, we have

$$\frac{\partial J(W, b)}{\partial W^l} = \frac{\partial J(W, b)}{\partial z^l} \frac{\partial z^l}{\partial W^l} = \delta^l \text{rot } 180(a^{l-1}).$$

For b, since δ^l is a three-dimensional tensor, and b is just a vector, it cannot be directly equal to δ^l like DNN. The usual practice is to sum the terms of each sub-matrix of δ^l separately to get an error vector, which is the gradient of b,

$$\frac{\partial J(W, b)}{\partial b^l} = \sum_{u,v} (\delta^l)_{u,v}.$$

5.1.3   Recurrent Neural Network and Long Short-Term Network

(1) RNN's loop structure

RNN is a network that contains loops, allowing information to be persisted. A, which is a module of the neural network, is reading an input x_i and outputting a value h_i. Looping allows information to be passed from the current step to the next step [2]. RNN can be thought of as multiple assignments of the same neural network, and each neural network module passes the message to the next one (see Fig. 5.4).

The chained features reveal that RNN is essentially related to sequences and lists. They are the most natural neural network architecture for this type of data. And RNN has also been applied in the past few years. The application of RNN has achieved some success in speech recognition, language modeling, translation, picture description, and other issues, and this list is still growing. The key to these successful

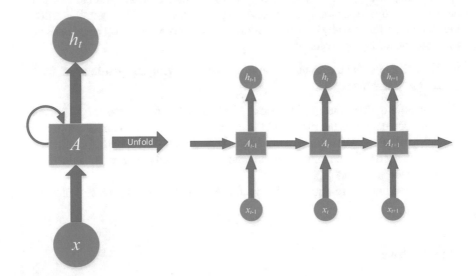

Fig. 5.4 RNN structure and its post-expansion chain characteristics

applications is the use of LSTM, a special RNN that performs better on many tasks than the standard RNN.

(2) **Long-term dependencies**

One of the key points of RNN is that they can be used to connect previous information to current tasks, such as using past video segments to infer the understanding of the current segment. But RNN has many other factors to depend on. Sometimes, we just need to know the previous information to perform the current task. For example, we have a language model for predicting the next word based on previous words. If we try to predict the last word of "the clouds are in the sky," we do not need any other context—so the next word should obviously be sky. In such scenarios, the margin between the relevant information and the predicted word position is very small, and RNN can learn to use the previous information.

But there will also be more complex scenarios. Suppose we try to predict the last word of "I grew up in France ... I spoke fluent French". Current information suggests that the next word may be the name of a language, but if we need to figure out what language it is, we need the context of France, which was mentioned earlier, far from the current location. This means that the gap between the relevant information and the current predicted position must be considerable. Unfortunately, RNN loses the ability to learn to connect information so far as this gap increases. Therefore, researchers have found a long short-term, LSTM model to solve such problems.

(3) **Initial knowledge of LSTM network**

The long short term network (LSTM) is a special type of RNN, which can learn to rely on information for a long time. LSTM was proposed by Hochreiter and Schmidhuber [3], and recently improved and promoted by Alex Graves. In many problems, LSTM has achieved considerable success and has been widely used.

LSTM avoids long-term dependency problems through deliberate design. Keep in mind that long-term information is the default behavior of LSTM in practice, rather than the ability to acquire it at great cost. All RNNs have a chain form of repetitive neural network modules. In standard RNN, this duplicate module has only one very simple structure, such as a tanh layer. LSTM is the same structure, but repetitive modules have a different structure. Unlike a single neural network layer, there are four layers that interact in a very special way, as is shown in Fig. 5.5.

First, let us familiarize ourselves with the icons of various elements used in the diagram (see Fig. 5.6).

In the above illustration, each black line transmits an entire vector from the output of one node to the input of other nodes. Pink circles represent pointwise operations, such as the sum of vectors, while yellow matrices are the learnt layers of neural networks. The joined lines represent the connection of vectors, and the separated lines represent that the content is copied and then distributed to different locations.

The key to LSTM is the cell state, and the horizontal line runs through the top of the graph. The cell state is similar to that of the conveyor belt. It runs directly across the chain, with only a few linear interactions. It is easy to keep the information going.

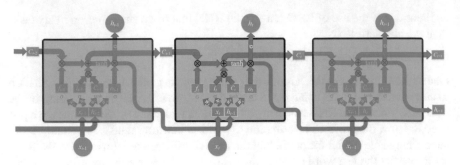

Fig. 5.5 Repetitive module in LSTM consists of four interactive layers

Vector   Concatenate   Copy
Transfer

Fig. 5.6 Icons in LSTM

LSTM has the ability to remove or increase information to the cellular state by carefully designed structures called gates. Gates are a way to allow information to pass selectively. They include a sigmoid neural network layer and a pointwise multiplication operation. The sigmoid layer outputs values between 0 and 1, describing how much each part can pass. 0 stands for "no quantity is allowed to pass," and 1 means "any quantity is allowed to pass." LSTM has three gates to protect and control cell status.

(4) **Understanding LSTM step by step**

The first step in our LSTM is to decide what information we will discard from the cell state. This decision is made through a layer called forgetting doors. The gate reads $h\_\{t - 1\}$ and x_t and outputs a value between 0 and 1 to each number in the cell state $C\_\{t - 1\}$. 1 means "complete reservation" and 0 means "complete abandonment."

Let us go back to the example of the language model to predict the next word based on what we have seen. In this problem, the cell state may contain the category of the current subject, so the correct pronoun can be selected. When we see new pronouns, we want to forget the old ones (see Fig. 5.7).

The next step is to determine what new information is stored in the cell state. There are two parts. First, the sigmoid layer is called the "input gate layer" to determine what values we will update. Then, a tanh layer creates a new candidate value vector, tilde$\{C\}\_t$, which is added to the state. Next, we talk about these two pieces of information to generate status updates.

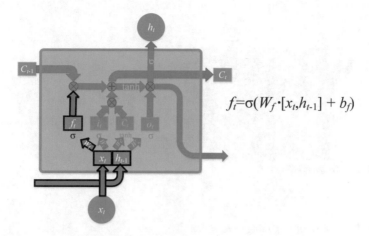

$$f_t = \sigma(W_f \cdot [x_t, h_{t-1}] + b_f)$$

Fig. 5.7 Process of deciding to discard information

In the example of our language model, we want to add new pronoun categories to the cellular state to replace the old ones that need to be forgotten (see Fig. 5.8).

Now, it is time to update the old cell state, $C\_\{t-1\}$ to C_t. The previous steps have decided what to do, and now we are actually going to do it.

We multiply the old state by f_t and discard the information we determined we needed to discard. Then, add $i\_t *$ tilde $\{C\}\_t$. This is the new candidate value, which varies according to the degree to which we decide to update each state.

In the case of language models, this is where we actually discard the category information of old pronouns and add new information based on the goals we set earlier.

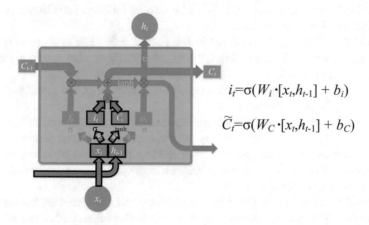

$$i_t = \sigma(W_i \cdot [x_t, h_{t-1}] + b_i)$$

$$\widetilde{C}_t = \sigma(W_C \cdot [x_t, h_{t-1}] + b_C)$$

Fig. 5.8 Process of identifying updated information

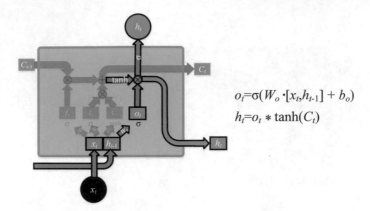

$$o_t = \sigma(W_o \cdot [x_t, h_{t-1}] + b_o)$$
$$h_t = o_t * \tanh(C_t)$$

Fig. 5.9 The process of outputting information

Ultimately, we need to determine what value to output. This output will be based on our cell status, but it is also a filtered version. First, we run a sigmoid layer to determine which part of the cell's state will be exported. Next, we process the cell state through tanh (to get a value between -1 and 1) and multiply it with the output of the sigmoid gate. Eventually, we only output the part of the output we determined.

In the case of a language model, because he sees a pronoun, he may need to output information related to a verb. For example, it is possible to output whether the pronoun is singular or negative, so if it is a verb, we also know the morpheme changes that the verb needs (see Fig. 5.9).

(5) Variants of LSTM

So far, we have been introducing normal LSTM. But not all LSTMs look the same. In fact, almost all papers containing LSTM use small variants. The difference is very small, but it is worth mentioning.

One of the LSTM variants of manifolds, proposed by Gers et al. (2000), added "peephole connection." That is to say, we let the portal layer also accept the input of cell state (see Fig. 5.10).

In the above illustration, we added peepholes to each door, but many papers will add some peepholes instead of all.

Another variant is forgetting and entering doors by using coupled. Unlike before, which separately identifies what forgets and what new information needs to be added, here is a decision to be made together. We just forget when we are going to type in the current location. We just enter new values to those states where we have forgotten the old information (see Fig. 5.11).

Another major variant is gated recurrent unit (GRU). It combines the forgetting gate and the input gate into a single update gate. It also mixes cell state and hidden state with other changes. The final model is simpler than the standard LSTM model and is a very popular variant (see Fig. 5.12).

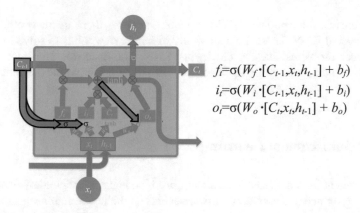

$$f_t = \sigma(W_f \cdot [C_{t-1}, x_t, h_{t-1}] + b_f)$$
$$i_t = \sigma(W_i \cdot [C_{t-1}, x_t, h_{t-1}] + b_i)$$
$$o_t = \sigma(W_o \cdot [C_t, x_t, h_{t-1}] + b_o)$$

Fig. 5.10 Peephole connection

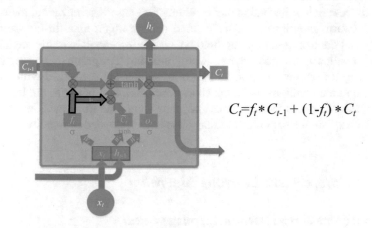

$$C_t = f_t * C_{t-1} + (1 - f_t) * C_t$$

Fig. 5.11 Coupled forget gate and input gate

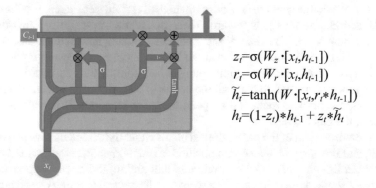

$$z_t = \sigma(W_z \cdot [x_t, h_{t-1}])$$
$$r_t = \sigma(W_r \cdot [x_t, h_{t-1}])$$
$$\widetilde{h}_t = \tanh(W \cdot [x_t, r_t * h_{t-1}])$$
$$h_t = (1 - z_t) * h_{t-1} + z_t * \widetilde{h}_t$$

Fig. 5.12 Gated recurrent unit

Here are just some popular LSTM variants. Of course, there are many others, such as depth-gated RNN. There are also totally different viewpoints to solve long-term dependence problems, such as clockwork RNN. The distortion and improvement of LSTM are also constantly updated. We can pay attention to the latest academic publications.

5.2 Reinforcement Learning

Reinforcement learning (RL) is an area of machine learning inspired by behaviorist psychology, concerned with how software agents ought to take actions in an environment so as to maximize some notion of cumulative reward. The problem, due to its generality, is studied in many other disciplines, such as game theory, control theory, operations research, information theory, simulation-based optimization, multi-agent systems, swarm intelligence, statistics, and genetic algorithms. In the operations research and control literature, the field where reinforcement learning methods are studying is called approximate dynamic programming. The problem has been studied in the theory of optimal control, though most studies are concerned with the existence of optimal solutions and their characterization, and not with the learning or approximation aspects. In economics and game theory, reinforcement learning may be used to explain how equilibrium may arise under bounded rationality.

5.2.1 Reinforcement Learning Algorithm

(1) Description of reinforcement learning process

There are five elements in reinforcement learning: Agent, environment, state, reward, and action. We define agent in an environment. In a certain state, agent has the ability to make action, and thus gets reward. These behaviors transform the environment and create new states, and then agents continue to make new actions and get new rewards, and so on. We define a rule about how to choose an action as a policy.

In the case of <break out> game, players can control the red board under the screen to rebound the ball to clear all the bricks in the half of the screen. Every time the ball hits the brick to make it disappear, the score increases. When all the bricks disappear, the game wins.

In the environment of the game, the red board under the screen that we can control is agent, and the agent is in a certain position is a state, which is represented as four states in the figure. Agent makes a rebound ball action, the brick disappears, then agent gets a score, namely reward. In a word, reinforcement learning can be regarded as the process of maximizing reward by agent in the current environment.

(2) **Markov decision process representation of reinforcement learning**

Markov decision process is widely used to describe reinforcement learning process in mathematical and computational languages.

Markov process is a kind of stochastic process with Markov property. That is, $P[S_{t+1}|S_t] = P[S_{t+1}|S_t...S_1]$, S_{t+1} is only related to the former state S_t, but not to S_{t-1} and the further state.

1. **Markov reward process (MRP)**

Markov reward process can be represented by $<S, P, R, \gamma>$. Among them:

S: Represents the state collection.

P: Represents the state transition matrix from state S_t to state S_{t+1}.

R: Represents reward function, defined as $R_s = E[R_{t+1}|S_t = s]$.

γ: Represents the discount rate for learning over time. Discount factor.

Under a policy, the state in which the execution action produces a sequence is defined as the cumulative discount return G_t, which is the sum of the feedback from each step of the conversion. The formula is expressed as,

$$G_t = R_{t+1} + \gamma R_{t+2} + \gamma^2 R_{t+3} + \cdots$$

Because the strategy is random, the cumulative return of discount is also random, so we cannot directly use G_t to describe the value function of state value, but its expectation is a certain value. Therefore, the value function $V(s)$ is defined as,

$$V(s) = E[G_t|S_t - s].$$

According to the basic principle of Bellman equation, the value of the current state is only related to the value of the next step and the current reward. So, we can get the formula,

$$V(s) = E[R_{t+1} + \gamma V(S_{t+1})|S_t = s] = R_s + \gamma \sum_{s' \in S} P_{s,s'} V(s').$$

This shows that the value function $V(s)$ can be calculated by iteration to obtain the value of each state. Bellman equation's derivation process is as follows:

$$
\begin{aligned}
V(s) &= E[G_t|S_t = s] \\
&= E[R_{t+1} + \gamma R_{t+2} + \gamma^2 R_{t+3} + \cdots |S_t = s] \\
&= E[R_{t+1} + \gamma(R_{t+2} + \gamma R_{t+3} + \cdots)|S_t = s] \\
&= E[R_{t+1} + \gamma G_{t+1}|S_t = s] \\
&= E[R_{t+1} + \gamma V(S_{t+1})|S_t = s] \\
&= R_s + \gamma \sum_{s \in S} P_{s,s'} V(s').
\end{aligned}
$$

2. **Markov decision process (MDP)**

Markov decision process can be represented by $<S, P, R, A, \gamma>$. Compared with Markov reward process, MDP adds A-action set and strategy π, and has two value-related functions: state–action value function $Q(s, a)$ and value function $V(s)$. Therefore, elements in MDP are defined as,

S: Represents the state collection.

A: Represents the action collection.

P: Represents a three-dimensional state transition matrix,

$$P_{s,s'}^a = P[S_{t+1} = s' | A_t = a, S_t = s].$$

R: Represents reward function. Maybe it is a $S \times A \mapsto R$; or maybe it is

$$R: S \mapsto R, \ R_s^a = E[R_{t+1} | A_t = a, S_t = s].$$

γ: Represents the discount rate for learning over time.

In MDP, agents make action choices at each step according to a certain probability. Therefore, we define the formula of policy as $\pi(a|s) = P[A_t = a | S_t = s]$, which represents the probability distribution of executing different actions a in state s.

In order to calculate the value function $V_\pi(s)$ after adding strategy, the state–action value function $Q_\pi(s, a)$ should be calculated first. $Q_\pi(s, a)$ is in a certain state, considering the value of different states, the formula is expressed as,

$$
\begin{aligned}
Q_\pi(s, a) &= E_\pi[R_{t+1} + \gamma R_{t+2} + \gamma^2 R_{t+3} + \cdots | A_t = a, S_t = s] \\
&= E_\pi[R_{t+1} + \gamma Q_\pi(S_{t+1}, A_{t+1}) | A_t = a, S_t = s] \\
&= R_s^a + \gamma \sum_{s' \in S} P_{s,s'}^a V_\pi(s').
\end{aligned}
$$

So, the value function v of state s and the sum of different actions are expressed as follows:

$$
\begin{aligned}
V_\pi(s) &= \sum_{a \in A} \pi(a|s) Q_\pi(s, a) \\
&= \sum_{a \in A} \pi(a|s) (R_s^a + \gamma \sum_{s' \in S} P_{s,s'}^a V_\pi(s')).
\end{aligned}
$$

So, $V_\pi = R_\pi + \gamma P_\pi V_\pi$, so we can get: $V_\pi = (1 - \gamma P_\pi) R_\pi$. Therefore, under the given strategy, the value function $Q(s, a)$ and $V(s)$ of state–action can be obtained.

3. **Optimal value function and optimal state–action value function**

Further, in order to achieve the ultimate goal of reinforcement learning, Goal, that is, to get the greatest reward value. The optimal value function and the optimal state–action value function are defined, respectively. For an MDP:

There exists an optimal strategy π_* such that for $\forall \pi$, $\pi_* \geq \pi$.

The value function corresponding to all optimal strategies is equal to the optimal value, i.e., $V_{\pi_*}(s) = V_*(s)$.

The state–action value function corresponding to all optimal strategies is equal to the optimal state–action value function, i.e., $Q_{\pi*} = Q_*(s, a)$.

Furthermore, Bellman's optimal equation is expressed as,

$$V_*(s) = \max_a Q_*(s, a),$$

$$Q_*(s, a) = R_s^a + \gamma \sum_{s' \in S} P_{s,s'}^a V_*(s').$$

(3) Two most commonly used iteration methods

The two most commonly used iterations in reinforcement learning are value iteration and policy iteration.

(A) Policy iteration

In Policy iteration algorithms, you start with a random policy, then find the value function of that policy (policy evaluation step), then find a new (improved) policy based on the previous value function, and so on. In this process, each policy is guaranteed to be a strict improvement over the previous one (unless it is already optimal). Given a policy, its value function can be obtained using the Bellman operator.

1. Initialization

Initialize $v(s)$ and $\pi(s)$ of all states (initialized as random strategy).

2. Policy evaluation

The current strategy is evaluated with the current $v(s)$, and the $v(s)$ of each state is calculated. The state value function $v(s)$ is not trained until $v(s)$ converges.

3. Policy improvement

Now that the evaluation function $v(s)$ of the current strategy has been obtained in the previous step, and we can use this evaluation function to improve the strategy.

For each possible action a, the expected value of the next state reached after taking this action is calculated at each state s. See which action can reach the maximum expected value function of the state, then select this action. In this way, $\pi(s)$ is updated, and then the above two and three steps are repeated until $v(s)$ and $\pi(s)$ converge.

(B) Value iteration

In value iteration, you start with a random value function and then find a new (improved) value function in an iterative process, until reaching theoptimal value function. Notice that you can derive easily the optimal policy from the optimal value function. This process is based on the optimality Bellman operator.

1. Initialization

Initialize $v(s)$ of all states.

2. Finding optimal value function

Notice the max in the pseudocode. For each current state s, for each possible action a, calculate the expected value of the next state to be reached after taking this action. To see which action can reach the maximum expected value function of the state, we take the maximum expected value function as the value function $v(s)$ of the current state to cycle through this step, until the value function converges, we can get the optimal value function.

3. Policy improvement

Using the optimal value function and state transition probability obtained from the above steps, we can calculate the optimal action that each state should take, which is deterministic.

Now that the evaluation function $v(s)$ of the current strategy has been obtained in the previous step, we can use this evaluation function to improve the strategy.

The differences and connections are:

1. Policy evaluation, the second step of policy iteration, is very similar to finding optimal value function, the second part of value iteration, except that the latter uses max operation, the former does not have max. So, the latter can get optimal value function, while the former cannot get optimal function.
2. The convergence speed of strategy iteration is faster. When the state space is small, it is better to choose strategy iteration method. When the state space is large, the computational complexity of value iteration is smaller.

5.2.2 Deep Reinforcement Learning

Deep learning has good perception ability, especially the emergence of CNN algorithm, which can extract the feature state of the image more effectively. Reinforcement learning can make effective decisions in the process of interaction with the environment. Therefore, deep reinforcement learning combines deep learning's perception and reinforcement learning's decision-making ability, which can better solve many practical problems.

(1) **Deep reinforcement learning based on value function-deep Q-network (DQN)**

Manih et al. [4, 5] combined convolutional neural network with Q-learning [6] algorithm in traditional RL, and proposed a deep Q-network (DQN) model. This model is a pioneering work in the field of DRL, which is used to deal with visual perception-based control tasks.

The input of DQN model is four preprocessed images nearest to the current time. The input is transformed nonlinearly by three convolution layers and two full connection layers, and finally, the Q value of each action is generated in the output layer.

In order to alleviate the instability of the nonlinear network representation function, DQN mainly improves the traditional Q-learning algorithm in three ways.

1. DQN uses experience replay to process the transferred sample $e_t = (s_t, a_t, r_t, s_{t+1})$ online. At each time step t, the transfer samples obtained from the interaction between agent and environment are stored in the playback memory unit $D = \{e_1, \ldots, e_t\}$. During training, small batches of transfer samples are randomly extracted from θ, and network parameters are updated by stochastic gradient descent (SGD) algorithm. When training deep network, it is usually required that the samples are independent of each other. This random sampling method greatly reduces the correlation between samples and improves the stability of the algorithm.

2. In addition to using the deep convolution network to approximate the current value function, DQN also uses another network to generate the target Q value. Specifically, $Q(s, a|\theta_i)$ represents the output of the current value network, which is used to evaluate the value function of the current state–action pair; $Q(s, a|\theta_i^-)$ represents the output of the target value network, and generally uses $Y_i = r + \gamma \max_{a'} Q(s', a'|\theta_i^-)$ approximation to represent the optimization objective of the value function, that is, the target Q value. The parameter θ of the current value network is updated in real time. After N iterations, the parameters of the current value network are copied to the target value network. The network parameters are updated by minimizing the mean square error between the current Q value and the target Q value. The error function is,

$$L(\theta_i) = E_{s,a,r,s'}[(Y_i - Q(s, a|\theta_i))^2].$$

The following gradients are obtained by calculating the partial derivatives of the parameter θ,

$$\nabla_{\theta_i} L(\theta_i) = E_{s,a,r,s'}[(Y_i - Q(s, a|\theta_i))\nabla_{\theta_i} Q(s, a|\theta_i)].$$

After introducing the target value network, the target Q value remains unchanged for a period of time, which reduces the correlation between the current Q value and the target Q value to a certain extent, and improves the stability of the algorithm.

3. DQN reduces the reward and error items to a limited range, guarantees that the Q and gradient values are within a reasonable range, and improves the stability of the algorithm. Experiments show that DQN performs at a competitive level comparable to human players in solving complex problems such as injecting Atari 2600 games into real environments. Even in some less difficult non-strategic games, DQN outperforms experienced human players. In solving all kinds of DRL tasks based on visual perception, DQN uses the same set of network model, parameter setting, and training algorithm, which fully demonstrates that DQN method has strong adaptability and versatility.

(2) **Deep reinforcement learning based on strategic gradient**

Strategic gradient is a commonly used strategy optimization algorithm. It updates the strategy parameters by constantly calculating the gradient of the total reward on the strategy parameters, and finally converges to the optimal strategy [7]. Therefore, when solving DRL problems, we can use the parameter depth neural network to parameterize the representation strategy, and use the strategy gradient method to optimize the strategy. It is worth noting that when solving DRL problems, the first choice is to adopt a strategy gradient-based algorithm. The reason is that it can directly optimize the expected total reward of the strategy, and directly search the optimal strategy in the strategy space in the end-to-end way, eliminating the tedious intermediate links. Therefore, compared with DQN and its improved model, the DRL method based on policy gradient has a wider scope of application and better effect of strategy optimization.

Strategic gradient method is a method that directly uses approximators to approximate and optimize strategies, and ultimately obtains the optimal strategy. This method optimizes the expected total reward of the strategy,

$$\max_{\theta} E[R|\pi_{\theta}],$$

where $R = \sum_{t=0}^{T-1} r_t$ represents the sum of rewards received in a scenario. The most common idea of strategy gradient is to increase the probability of higher total reward scenarios. The specific process of the strategy gradient method is as follows:

Assume that the state, action, and reward trajectory of a complete plot is, $\tau = (s_0, a_0, r_0, s_1, a_1, r_1, \ldots, s_{T-1}, a_{T-1}, r_{T-1}, s_T)$. Then, the policy gradient is expressed in the following form,

$$g = R\nabla_{\theta} \sum_{t=0}^{T-1} \log \pi(a_t|s_t; \theta).$$

Using this gradient to adjust the strategy parameters, we have

$$\theta \leftarrow \theta + ag,$$

where α is the learning rate and controls the rate at which the policy parameters are updated. The $\nabla_{\theta} \sum_{t=0}^{T-1} \log \pi(a_t|s_t; \theta)$ gradient term indicates the direction in which the probability of occurrence of the trajectory τ can be increased. After multiplying the score function R, the higher the total reward in a single episode, the more "forced" the probability density. That is, if a lot of trajectories of total rewards are collected, the above-mentioned training process will cause the probability density to move toward the higher trajectory of the total prize, thereby maximizing the probability of occurrence of the high prize trajectory τ.

In some cases, however, the total reward R for each scenario is not negative, and the value of all gradients G is greater than or equal to zero. At this time, every trajectory

τ encountered in the training process will make the probability density "close" in the positive direction, which greatly slows down the learning speed. This will make the variance of the gradient g large. Therefore, some standardized operation can be used to reduce the variance of gradient g for R. This technique improves the probability of trajectory τ with larger total reward R and reduces the probability of trajectory τ with smaller total reward R. According to the above ideas, Williams [8] proposed reinforce algorithm, which unifies the form of strategy gradient as follows,

$$g = \nabla_\theta \sum_{t=0}^{T-1} \log \pi(a_t|s_t; \theta)(R - b),$$

where b is a baseline related to the current trajectory τ and is usually set as an expected estimate of R in order to reduce the variance of R. It can be seen that the more R exceeds the base b, the greater the probability of the corresponding trajectory τ being selected. Therefore, in large-scale DRL tasks, deep neural network parametric representation strategy can be used, and the traditional strategy gradient method can be used to solve the optimal strategy.

In addition, another idea of optimization strategy is to increase the probability of "good" action. In RL, the dominant function is usually used to evaluate the performance of the action, so the dominant function term can be used to construct the strategy gradient,

$$g = \nabla_\theta \sum_{t=0}^{T-1} \hat{A}_t \log \pi(a_t|s_t; \theta),$$

where $\hat{A}_t$ is an estimate of the state action's (s_t, a_t)-dominant function, usually constructed as follows,

$$\hat{A}_t^\gamma = r_t + \gamma r_{t+1} + \gamma^2 r_{t+2} + \cdots - V(s_t).$$

Among them, $\gamma \in [0, 1]$ represents the discount factor. At this time, the sum of rewards with discounts $r_t + \gamma r_{t+1} + \gamma^2 r_{t+2} + \dots$ is equivalent to R in $g = \nabla_\theta \sum_{t=0}^{T-1} \log \pi(a_t|s_t; \theta)(R - b)$, and the state value function $V(s_t)$ with discounts is equivalent to the benchmark b. When $\hat{A}_t^\gamma > 0$, the probability of the corresponding action being selected will be increased, while when $\hat{A}_t^\gamma < 0$, the probability of the corresponding action being selected will be reduced.

In addition, Hafner and Riedmiller [9] used the value function to estimate the bonus sum with discount, which further reduced the variance of the gradient term. At this point, a step-truncated $\hat{A}_t^\gamma$ is expressed as,

$$\hat{A}_t^\gamma = r_t + \gamma V(s_{t+1}) - V(s_t).$$

Similarly, the two-step truncated $\hat{A}_t$ is expressed as,

$$\hat{A}_t^{\gamma} = r_t + \gamma r_{t+1} + \gamma^2 V(s_{t+2}) - V(s_t).$$

However, the use of value function to estimate the sum of rewards with discounts will also produce some estimation bias. In order to reduce the variance and ensure the smaller deviation, Schulman et al. [10] proposed generalized advantage function,

$$\hat{A}_t^{\gamma} = \delta_t + (\gamma\lambda)\delta_{t+1} + (\gamma\lambda)^2\delta_{t+2} + \cdots + (\gamma\lambda)^{T-t-1}\delta_{T-1}.$$

Among them, $\delta_t = r_t + \gamma V(s_{t+1}) - V(s_t)$. λ is a regulatory factor with a range of $0 < \lambda < 1$. When λ is close to 0, $\hat{A}_t^{\gamma}$ is low variance and high error; when λ is close to 1, $\hat{A}_t^{\gamma}$ is high variance and low error. The disadvantage of the strategy gradient method based on generalized dominance function is that it is difficult to determine a reasonable step parameter α to ensure the stability of learning in the process of using $\theta \leftarrow \theta + ag$ global optimization strategy. To solve this problem, Schulman et al. [11] proposed a strategy optimization (TRPO) method which is called regional trust. The key idea of TRPO is to restrict the KL difference between the old and the new strategies on the same batch of data, so as to avoid parameter updating steps that lead to too much change in the strategy. In order to extend the application scope to DRL tasks in large-scale state space, TRPO algorithm uses deep neural network to parameterize the strategy and achieves end-to-end control when only the original input image is received. Experiments show that TRPO performs well in robot control and Atari 2600 game tasks in a series of 2D scenarios. Since then, Schulman et al. [10] have tried to combine the generalized dominance function with TRPO method, and made breakthroughs in a series of robot control tasks in 3D scenes.

In addition, another research direction of deep strategy gradient method is to promote strategy search by adding additional manual supervision. For example, the famous AlphaGo robot [11], first uses supervised learning to predict human walking behavior from the chess game of human experts, and then uses the strategy gradient method to fine-tune the strategy parameters aiming at the real goal of winning the Go game. However, there is a lack of monitoring data in some tasks, such as robot control in real-world scenarios, and the guided policy search method can be used to monitor the process of strategy search. In the real scene where only the original input signal is received, the guided strategy search achieves the manipulation of the robot [12].

(3) **Deep reinforcement learning based on search and supervision**

In addition to value-based DRL and policy gradient-based DRL, additional manual supervision can be added to promote the process of policy search, which is the key idea of DRL based on search and supervision. As a classical heuristic strategy search method, Monte Carlo tree search (MCTS) [13] is widely used in action planning in game problems. Therefore, in the DRL method based on search and supervision, policy search is usually accomplished through MCTS. The AlphaGo algorithm introduced in this section combines deep neural network with MCTS, and achieves remarkable achievements [11].

In the field of AI, owing to the huge state space and the difficulty in accurately evaluating the expected layout and walking, it has always been considered the most challenging problem to develop an agent who can master Go game. Until Silver et al. combined CNN with MCTS, proposed a Go algorithm called AlphaGo, which solved this problem to a certain extent. There are two main ideas of AlphaGo: (1) using MCTS to approximate the value function of each state; (2) using CNN based on value function to evaluate the current layout and walk of chessboard. AlphaGo's complete learning system consists of the following four parts:

1. Policy network. It also monitors the learning strategy network and RL strategy network. The role of the strategy network is to predict and sample the next move according to the current situation.
2. Rollout policy. The goal is to predict the next step, but the prediction speed is 1000 times faster than that of the strategic network.
3. Value network. According to the current situation, the probability of winning is estimated.
4. MCTS. The strategy network, roller network, and valuation network are integrated into the strategy search process to form a complete system.

Firstly, in the first stage of training AlphaGo, the strategy network P_σ of supervised learning is trained through tagged game data on the Go game server KGS. The ultimate goal is to simulate the human players' walking pattern a under the current chessboard state s,

$$\Delta\sigma \propto \frac{\partial \log P_\sigma(a|s)}{\partial \sigma},$$

where σ represents the parameters of the policy network that supervises learning. The policy network is a 13-layer deep convolutional network. The specific training method is the gradient descent method. After the training, all the input features were used on the test set, and the accuracy rate of the human professional chess piece movement was predicted to be 57.0%. When only the board position and historical travel record are used as input, and the accuracy of the prediction is also 55.7%. In addition, using local feature matching and linear regression to train the wheel strategy network, the accuracy of predicting human professional player moves is 24.2%.

Secondly, in the second stage of training AlphaGo, the strategic gradient method is used to train the RL's policy network P_ρ to further improve the walking ability of the strategy network, and finally maximize the expected reward of the whole game,

$$\Delta\rho \propto \frac{\partial \log P_\rho(a_t|s_t)}{\partial \rho} z_t,$$

where ρ is the parameter of the RL's policy network, and z_t is the final gain of a game of chess. The win is $+1$ and the negative is -1. The specific training method is: randomly select the policy network of the previous iteration round and the current strategy. The network P_ρ plays against each other and uses the policy gradient method to update the parameters, and finally obtains an enhanced policy network. P_ρ is identical in structure to P_σ. After training, the enhanced P_ρ has a winning percentage of over 80% against P_σ and an 85% win against Pachi (a Go algorithm).

Then, in the third phase of training AlphaGo, the main focus is on the valuation of the current situation. During training, the evaluation network is trained by minimizing the mean square error between the output network $v_\theta(s)$ and the benefit z,

$$\Delta\theta \propto \frac{\partial v_\theta(s)}{\partial\theta}(z - v_\theta(s)).$$

The network structure used in the valuation network is similar to the strategy network. The difference between the two is that the evaluation network only outputs a single predicted value v at the output layer, which is used to estimate the probability of black or white winning, while the strategy network. The output is a probability distribution of possible walking actions.

Finally, AlphaGo combines the MCTS algorithm with the strategy network and the valuation network, and selects the move action through advanced search. At each time step t, select a move action a_t from state s_t,

$$a_t = \arg\max_a(Q(s_t, a) + u(s_t, a)).$$

Among them, $u(s_t, a)$ represents an additional reward, with the aim of maximizing the value of the walking sub-action on the premise of encouraging exploration,

$$u(s, a) \propto \frac{P(s, a)}{1 + N(s, a)},$$

where $P(s, a) = P_\sigma(a|s)$ denotes the output of the policy network as the prior probability and $N(s, a)$ denotes the number of accesses of the state–action pair. $u(s, a)$ is proportional to the prior probability and inversely proportional to the number of visits. Then, when a leaf node S_L is reached after traversing the L step, the evaluation network $v_\theta(s_L)$ and the wheel strategy network are evaluated to obtain the value of the leaf node,

$$V(s_L) = (1 - \lambda)v_\theta(s_L) + \lambda z_L,$$

where z_L represents the reward obtained when the game is terminated. Then, update the number of accesses of the state–action pair and the corresponding action value,

$$N(s, a) = \sum_{i=1}^{N} l(s, a, i),$$

$$Q(s, a) = \frac{1}{N(s, a)} \sum_{i=1}^{N} l(s, a, i)V(s_L^i),$$

where $l(s, a, i)$ is related to whether the state–action pair (s, a) is accessed in the ith simulation, specifically when the value is set to 1 when accessed, and when the value is not accessed, the value is set to 0; s_L^i indicates the leaf of the ith simulation state node. Once the search is complete, the agent selects the most visited move from the position of the root node.

After the training, AlphaGo defeated a European champion and a world-class champion player, which fully proved that the computer-based Go algorithm based on DRL algorithm has reached the level of human top players. The success of AlphaGo is a milestone for the development of general artificial intelligence.

(4) **Other DRL**

In addition to the most commonly used DRL algorithms mentioned above, some DRL algorithms which perform better on specific problems are derived for different practical application scenarios. For example, in some complex DRL tasks, it is inefficient to optimize strategies directly based on the ultimate goal. Hierarchical reinforcement learning (HRL) [14] can be used to decompose the final goal into multiple sub-tasks to learn hierarchical strategies, and to form effective global strategies by combining multiple sub-tasks.

In addition, in the traditional DRL method, each agent after training can only solve a single task. However, in some complex realistic scenarios, agent is required to be able to handle multiple tasks at the same time. At this time, multi-task learning and transfer learning are extremely important. In RL domain, Yi et al. [15] used Bayesian model to provide prior knowledge about new tasks. Li et al. [16] proposed regionalized policy representation to describe the behavior of agents in different task scenarios for some observable random multi-scene tasks. This method utilizes the clustering property contained in Dirichlet process to share training scenarios between similar tasks and transmit valuable information among different tasks. Compared with single-task learning mode, this multi-task RL method has achieved more outstanding performance in grid world navigation and multi-target classification tasks. Hu et al. [17] proposed a distribution-free probability density function for solving the problem. Fernandez and Veloso [18] used a mapping reflecting the action relationship of the agent between the current and past state, so that the strategies learned in the past can be used in a new task. Wang et al. concluded that migration in RL can be divided into two categories: behavior migration and knowledge migration, which are also widely used in multi-task DRL algorithm.

Furthermore, the decision-making ability of single-agent system is far from enough when facing some complex decision-making problems in real scenarios. For example, in the multi-player Atari 2600 game, there is a need for cooperation or competition among multiple decision-makers. Therefore, in a specific case, it is necessary to extend DRL model to a multi-agent system that cooperates, communicates and competes with each other among agents [19].

Finally, the performance of traditional DRL methods based on visual perception in solving higher-level cognitive-inspired tasks is far from that of human beings. That is to say, when solving some high-level DRL tasks, agents need not only strong perception ability, but also certain memory and reasoning ability to learn effective decision-making. Therefore, it is very important to endow existing DRL models with the ability of active memory and reasoning. In recent years, substantial progress has been made in the study of neural network models with external storage. Graves et al. [20] proposed a memory structure called Neural Turing Machines (NTM), which updates the parameters of memory structure and optimizes the content of memory by means of random gradient descent while reading and writing data. By adding NTM, the neural network model has the ability to accomplish some simple tasks such as replication, inversion, addition, and subtraction. It shows that the deep neural network model has the ability of preliminary memory and reasoning. Subsequently, Sukhbaatar et al. [21] proposed a memory network model based on NTM for question answering system and language modeling task, which further improved the long-term memory ability of the network. Therefore, adding these external memory modules to the existing DRL model can give the network some high-level abilities such as long-term memory, active cognition, reasoning, and so on. In addition, the development of cognitive neuroscience in recent years has also promoted the development of artificial intelligence to a certain extent. People are simulating the human brain's assistant learning system to construct an agent that can independently remember, learn, and make decisions.

5.3 Brother Learning

In the long history of human evolution, the appearance of characters is often regarded as an important symbol of civilization. The appearance of words means that knowledge and experience can be inherited. Later, humans can use the guidance of their predecessors' knowledge and experience to save the cost of learning, so that they can continuously acquire more skills than their predecessors, and lead other creatures on the path of evolution. In fact, the modern society also has the shadow of inheritance and evolution. In a family with many children, if the elder brother (sister) has better academic performance, the younger brother (sister) often has better academic performance. The reason is that an excellent elder brother (elder sister) can share his learning methods and unique skills with his younger brother in time, so that his younger brother can save the time cost of learning and more easily achieve good results. In this section, we call this kind of model that predecessors can share their knowledge or learning methods with future generations, so that they can learn more efficiently as "brother learning." In fact, in the process of researching artificial intelligence, we can also refer to this pattern in the evolution of human intelligence, that is, how to transfer the existing experience and knowledge of the model.

This section takes the classification model as an example to propose the main paradigm of brother learning (see Fig. 5.13).

For the general classification model, the training and learning process is based on fixed categories. However, in the actual application scenarios, with the business iteration, it is very common to continuously add new learning categories. At the same time, it is difficult to achieve sufficient and balanced data available for each

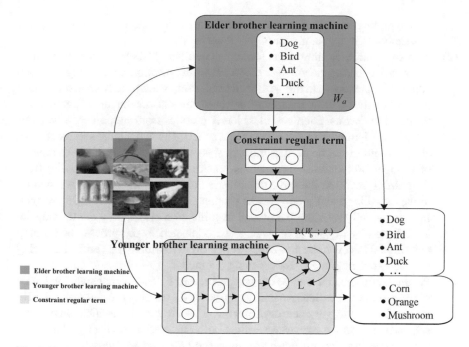

Fig. 5.13 Basic framework of brother learning

category. Therefore, how to remember the learned categories and recognize the new few sample categories better is a very important problem in machine learning field. That is to say, when the elder learning machine can recognize a group of basic categories, by sharing the parameters of the elder learning machine, the younger learning machine cannot only recognize the basic categories that the elder learning machine can recognize, but also consider additional new categories with only a small number of labels. The main stages of brother learning are shown in Fig. 5.13: (1) First, a classifier is trained on the inherent category by traditional supervised learning, and a fixed feature expression is learned, which is called elder brother learning machine; (2) At each training and testing node, this section trains a new classifier based on the meta-learning regularization matrix, which is equivalent to learning new knowledge by the younger brother's learning machine; (3) Combining the new category and the inherent category and optimizing the regular module in the previous stage, it can play the same role after combining the inherent classifier. That is, after incorporating the parameters of elder brother's learning machine, through optimization, younger brother's learning machine cannot only identify a group of basic categories, but also consider some additional new categories.

Each stage involved will be specifically unfolded as follows:

(1) Brother learning machine training stage: in this stage, there is no additional special operation. Given all the data and corresponding category labels of the inherent category classification, train an inherent category classifier and its

corresponding feature expression, and then get a basic classification model. Remember that the parameter of this stage is W_a.

(2) New knowledge learning stage: for a small number of labeled new category data set D, it is used for the younger brother learning machine in this stage to learn additional categories. In this section, a small sample learning framework is used to describe the process. Each time, select K new classes that are different from the existing classes. Each new class has N pictures from support set S (support set) and M pictures from query set Q (query set). The two picture sets mentioned above, S and Q can be used as training sets and verification sets for each node of younger brother learning machine. Each node learns a new classifier from the training set S. At the same time, the learning parameter W_b corresponding to the classifier usually only works on this node, which is called fast weight. In order to fully measure the overall classification effect, only the training set S of the new category is allowed to be contacted in the process of training algorithm. However, the new category is used to reinforce the verification set Q with category merging when verifying the model.

(3) Parameter optimization stage of younger brother's learning machine: the whole process of meta-learning is to iteratively repeat the training process mentioned in the previous stage, get a new classifier on the newly added classification training set, and perform performance verification on the verification set Q. The cross entropy loss function and the extra regular term $R(\theta)$ are used as the optimization objective function to learn the updating speed parameter W_b. Among them, θ is a meta-parameter, which can be effectively learned by gradient descent or recursive backpropagation algorithm.

5.4 Epiphany Learning

5.4.1 The Concept of Epiphany

Epiphany or insight refers to the sudden understanding of the truth. It is a small leaping thinking process completed in a moment near the boundary of the knowledge field. For some really difficult problems, after the completion of "The dress takes to loosen gradually and I am increasingly emaciated, no regretful plying at all, I am rather for her only distressed as I did. Hundreds and thousands of times, for her I searched in chaos" (ancient Chinese poem), I got the cognitive result that "Suddenly, I turned by chance, to where the lights were waning, and there she stood!" (ancient Chinese poem).

In Buddhism, there is the theory of "suddenly realizing into Buddha," which is relative to the process of gradually realizing. Epiphany is usually accomplished by inspiration. Buddhism believes that "in terms of time, it may be an instant." Psychologists have found that "epiphany must have a clear premise of thinking about the problem. At the same time, epiphany must go through long-term, serious and even hard thinking about this problem before it can appear." Epiphany often follows a stage

Fig. 5.14 Five dime
converted into a triangle

of trial and error, but it is by no means an exhaustive process of brute force trial and error. Epiphany is sudden, induced, and accidental. Modern psychologists also find that epiphany is related to people's non-memory subconscious.

In the real world, the concept of epiphany is broad. Limited to the consideration of space and content concentration, this paper only considers epiphany learning or insight learning. In psychological research, epiphany research is also a problem being explored. In this respect, the famous thought model is Gestalt model. This paper tries to make an exploration in the direction of "structure" of epiphany from the perspective of artificial intelligence.

Epiphany is one of the highest human wisdom. Epiphany is a kind of learning with leaps. It is not a normal learning, but a learning with a jump in cognitive session. Therefore, it requires a smart wisdom, which is one of the highest forms of wisdom.

Epiphany starts with a goal, such as eating bananas or going to the restroom, and then lacks at least one cognitive link. The first time you see money folded into a triangle (see Fig. 5.14) or go to the restroom to see a sign you have never seen (see Fig. 5.15), it usually starts with "daze" and then you suddenly realize it, and then at least one learning link is missing. Epiphany must take a second or two. So, for an epiphany, there is a pause before an epiphany.

5.4.2 Learning with Teachers, Learning without Teachers, and Epiphany Learning

Epiphany learning has the shadow of both supervised learning (with teacher learning) (see area A in Fig. 5.16) and unsupervised learning (without teacher learning) (see area B in Fig. 5.16). It is a new learning category different from supervised learning and unsupervised learning. There is a shadow of supervised learning because simple epiphany has a clear goal. For example, when going to the restroom, there is a

Fig. 5.15 Restroom sign

Fig. 5.16 Learning with
teachers, learning without
teachers, and epiphany
learning

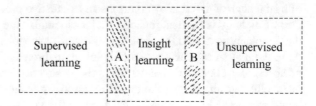

clear goal. When receiving the five dime banknote folded into a triangle shape, the
intuition in the brain is also five dime, which belongs to the category of supervised
learning. But for the Archimedes' buoyancy principle, before Archimedes found out
the principle and jumped out of the bathtub, he had no clear goal in his mind and
belonged to unsupervised learning.

Psychologists define epiphany as having new knowledge (such as Archimedes'
buoyancy principle). So we divided epiphany learning into two categories: one is
to generate huge new knowledge (more inclined to unsupervised learning), and the
other is to basically not generate new knowledge (supervised learning).

In the research of artificial intelligence, epiphany is rarely involved, and it is also
the front research direction at present.

5.4.3   Structured Description of Epiphany Learning

The boundary of epiphany jump may be linear or nonlinear. Linear means that the
thinking is in area I of $ax + b < 0$, while the goal is in area II of $ax + b > 0$ (see
Fig. 5.17). Nonlinearity refers to the region I where the thinking is in $g(x) < 0$, while
the goal is in region II where $g(x) > 0$ (see Fig. 5.18).

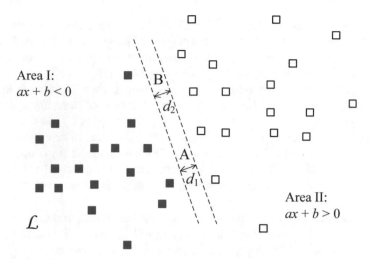

Fig. 5.17 Schematic diagram of linear boundary epiphany

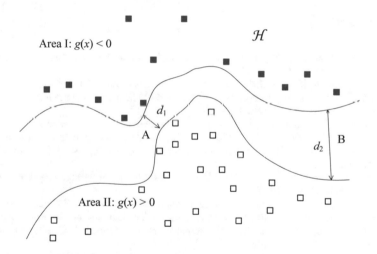

Fig. 5.18 Schematic diagram of nonlinear boundary epiphany learning

In this paper, the reason for dividing linear and fractional linear is that considering the boundary complexity of epiphany. The completion of epiphany is often a multiple "trial and error" process to complete epiphany, while the completion of epiphany is usually a simple trial and error at the "easiest" boundary. The diagram in Fig. 5.18 only shows the complexity of a boundary.

In the real world, the gap between epiphany jumps may be wide or narrow. For linear jumps, d should be equal everywhere (in Fig. 5.17, $d_1 = d_2$). For nonlinear jumps, d may not be equal everywhere (see Fig. 5.18, $d_1 < d_2$). People with different abilities may complete the epiphany "jump" in different places. For example, master-level people may be able to complete the epiphany "jump" when they reach a wider "gap" (for example, d_2). However, students may not be able to complete the epiphany "jump" until they reach a narrow "gap" (for example, d_1).

If we put the usual "learning" into enlightenment, then in people's learning, epiphany is more difficult than enlightenment. Because of the leaping nature of thinking, or even the leaping span of "rushing out of plight," not everyone can complete it. In the academic field, with the "leap" achievement, many scholars later became "elite" or "master," and in the history of human beings, many famous creative thinking are "sudden." Therefore, to study the creative thinking of machines, we cannot avoid using models and algorithms to simulate the framework of epiphany mechanism.

In this subsection, only a little attempt was made to identify and guess banknote. At present, the mechanism of epiphany learning of human brain is also an exploring problem in the fields of both psychology and artificial intelligence.

References

1. LeCun Y, Bottou L, Bengio Y (1998) Gradient-based learning applied to document recognition. Proc IEEE Conf 86(11):2278–2324
2. Tanaka G, Nakane R, Takeuchi T, Yamane T, Nakano D, Katayama Y, Hirose A (2020) Spatially arranged sparse recurrent neural networks for energy efficient associative memory. IEEE Trans Neural Netw Learn Syst 31(1):24–38
3. Hochreiter S, Schmidhuber J (1997) Long short-term memory. Neural Comput 9(8):1735–1780
4. Mnih V, Kavukcuoglu K, Silver D (2013) Playing atari with deep reinforcement learning. In: Proceedings of workshops at the 26th neural information processing systems 2013, Lake Tahoe, pp 201–220
5. Mnih V, Kavukcuoglu K, Silver D (2015) Human-level control through deep reinforcement learning. Nature 518(7540):529–533
6. Watkins CJ (1989) Learning from delayed rewards. Robot Auton Syst 15(4):233–235
7. Sutton RS, Mcallester DA, Singh SP (1999) Policy gradient methods for reinforcement learning with function approximation. In: Proceedings of the advances in neural information processing systems, Denver, pp 1057–1063
8. Williams RJ (1992) Simple statistical gradient-following algorithms for connectionist reinforcement learning. Mach Learn 8(3):229–256
9. Hafner R, Riedmiller M (2011) Reinforcement learning in feedback control. Mach Learn 84(1):137–169
10. Schulman J, Moritz P, Levine S (2015) High-dimensional continuous control using generalized advantage estimation. arXiv preprint arXiv:1506.02438
11. Schulman J, Levine S, Moritz P (2017) Trust region policy optimization. In: Proceedings of international conference on machine learning, Lugano, pp 1889–1897
12. Silver D, Huang A, Maddison CJ (2016) Mastering the game of go with deep neural networks and tree search. Nature 529(7587):484–489
13. Coulom R (2006) Efficient selectivity and backup operators in Monte-Carlo tree search. In: Proceedings of international conference on computers and games, Berlin, pp 72–83
14. Barto AG, Mahadevan S (2003) Recent advances in hierarchical reinforcement learning. Discrete Event Dyn Syst 13(4):341–379

15. Yi W, Park J, Kim J (2020) GeCo: classification restricted Boltzmann machine hardware for on-chip semisupervised learning and Bayesian inference. IEEE Trans Neural Netw Learn Syst 31(1):53–65
16. Li H, Liao X, Carin L (2009) Multi-task reinforcement learning in partially observable stochastic environments. J Mach Learn Res 10(3):1131–1186
17. Hu T, Guo Q, Li Z, Shen X, Sun H (2020) Distribution-free probability density forecast through deep neural networks. IEEE Trans Neural Netw Learn Syst 31(1):612–625
18. Fernandez F, Veloso M (2015) Probabilistic policy reuse in a reinforcement learning agent. In: Proceedings of international joint conference on autonomous agents and multiagent systems, Istanbul, pp 720–727
19. Tampuu A, Matiisen T, Kodelja D (2015) Multi-agent cooperation and competition with deep reinforcement learning. arXiv preprint arXiv:1511.08779
20. Graves A, Wayne G, Danihelka I (2014) Neural turing machines. arXiv preprint arXiv:1410.5401
21. Sukhbaatar S, Weston J, Fergus R (2015) End-to-end memory networks. In: Proceedings of the advances in neural information processing systems, Montreal, pp 2440–2448

Chapter 6
Social Network Data Mining and Knowledge Discovery

In Chap. 2, we mentioned that social computing involves disparate data including text, images, speech, multimedia, social media, spatial, temporal, spatiotemporal data, etc. It can be seen that the data forms in social computing are diverse. But we believe that the final form of data processing can be transformed into three forms: numerical data, text data, and image data. For example, temporal and spatiotemporal data extracted from Google trees can be converted into numerical form and stored in the database; voice conversations collected from social networking sites need to be converted into text form for subsequent analysis and processing; and video processing technology also draws on the relevant tools and methods of image processing. Therefore, in the following two sections, we select the text information and image information which are relatively difficult to deal with to elaborate.

6.1 Online Social Networks Text Processing Models

6.1.1 Information Extraction

Information extraction is to structure the information contained in the text into a form of organization. Generally, specific factual information is extracted from text. For example, extract details of terrorist incidents from news reports: time, place, target, weapons, etc. Usually, the extracted information is described in a structured form, which can be directly stored in the database for user query and further analysis.

(1) **Development history of information extraction research**

Information extraction research began in the mid-1960s, with the Linguistic String Project of New York University and the FRUMP Project of Yale University as representatives. Until the late 1980s, the research and application of information extraction gradually entered a prosperous period, which benefited from the convening of

© Tsinghua University Press 2020
X. Liang, *Social Computing with Artificial Intelligence*,
https://doi.org/10.1007/978-981-15-7760-4_6

Message Understanding Conference and MUC conference. From 1987 to 1997, MUC held seven sessions. MUC introduced scenario templates, named entity, coreferential relationship determination, template element filling, and template relationship tasks into information extraction system. Among them, named entity recognition tasks are mainly to identify and classify proper nouns and meaningful quantitative phrases in the text; coreferential relationship determination tasks are to identify the reference expressions of a given text and determine the coreferential relationship between these expressions; template element filling tasks are to identify all entities of a specific type and their attribute characteristics. A template relationship task is intended to determine relationships that are not related to a particular domain between entities. The MUC meeting also determined the main evaluation indicators for measuring the performance of the information extraction system: recall rate, accuracy rate, and the weighted geometric mean F index. All in all, the MUC meeting has greatly promoted the establishment and development of the research direction of information extraction. The various specifications of the information extraction tasks defined by MUC and the established evaluation system have become the de facto standard for information extraction research. Following the MUC, the automatic content extraction (ACE) organized by the National Institute of Standards and Technology (NIST) from 1999 to 2008 became another important international conference dedicated to information extraction research. Compared with MUC, ACE evaluation does not take a specific field or scenario of Nudi, based on a false report (there is no standard output in the system output) and false positives (not in the standard answer but in the system output). The evaluation system also evaluates the system's cross-document processing capabilities. In addition to MUC, ACE also has many international conferences, including multilingual entity task evaluation (MET), and document understanding conference (DUC) is equivalent to the international academic conferences related to information extraction, which greatly promotes the development of information extraction technology.

(2) **Key subtasks in information extraction subtasks**

(1) **Named entity recognition (NER)**

Named entity recognition is the basic work of information extraction, and it is also a technology for optimal practical value in information extraction research. Its task is to identify such things as person name, organization name, date, time, place, specific digital form, etc., and add corresponding annotation information to facilitate the follow-up work of information extraction [1].

NER methods are mainly divided into rule-based methods and statistics-based methods. In contrast, although the accuracy of rule-based method is better than that of statistics-based method, these rules often depend on specific language, domain, and text format. The compilation process is time-consuming and error-prone, which requires experienced linguists to complete. By contrast, statistical methods are trained with manual tagged corpus, which does not require extensive knowledge of computational linguistics and can be completed in a relatively short time. Therefore, statistical-based systems are more portable across domains.

(2) **Coreference resolution**

Coreference is a common linguistic phenomenon, which is usually divided into coreference and cocoreference. Coreference refers to the close semantic relationship between the current reference and the words, phrases, or sentences mentioned above, while cocoreference mainly refers to the same reference in the real world when multiple nouns (including pronouns and noun phrases) refer to the same reference. Coreference resolution can simplify and unify the presentation of entities, which can greatly improve the accuracy of information extraction results.

Early coreference resolution was mainly based on linguistic structure or domain knowledge, which was realized by artificial construction of resolution rules. The more influential research results mainly include the methods based on central theory and classification algorithm, multi-agent strategy based on restriction rules, etc.

Coreference resolution based on machine learning uses mature algorithms or models in machine learning to process pronouns and texts, and deconstructs them from the perspective of probability. For example, Carina and Ante [2] introduced the Implicit Role Linking method to detect the semantic roles that are not filled in the event and determine which directional information these roles may be understood in the context. The solution given in this paper is to embed the entity model into the unsupervised CR (coreference resolution) framework, and the resolution effect will be significantly improved. In addition, expectation maximization (EM) algorithm is used to transform coreference resolution into classification problem and cross-linguistic cocoreference resolution system based on maximum entropy classifier [3].

(3) **Relationship extraction**

Relation extraction refers to the acquisition of grammatical or semantic links between entities in text. Relation extraction is the key link in information extraction. Early relational extraction mainly used pattern matching method, and then, a lexicon-driven method was developed. At present, machine learning and ontology-based method are mainly used [4]. Machine-based learning is still classified according to text features in essence. For example, Datta and Das [5] proposed a multi-objective relation extraction method based on support vector machines. The classifier combines supervised and unsupervised information extraction methods. The results show that the method not only extracts entity relations from unmarked text, but also breaks away from dependence on domain knowledge to a certain extent.

6.1.2 Keywords Mining

Keyword is the main topic expressed in a document. When dealing with documents or sentences, extracting keywords is a very important task. In solving practical problems, a typical process of keyword extraction can be summarized as follows: original corpus → stop word filtering, word shape restoration, (possibly part of speech tagging), word segmentation →keyword extraction by algorithm → multi-algorithm result fusion, and final keywords can be generated. In this process, extracting keywords by algorithm is the most important step.

According to the existing algorithms, they can be roughly divided into a few categories: tf-idf, KEA, RAKE, and so on, and TextRank, based on mutual information and left–right entropy. The following is a brief description.

(1)   Term frequency-inverse document frequency (tf-idf)

tf-idf is a classic algorithm that is often used. tf stands for "keyword frequency" or "single-text vocabulary frequency," that is, the frequency of occurrence of a certain item in a document. The higher the tf, the stronger the ability to distinguish the attributes of the document content, the higher the weight; idf represents the "inverse text frequency index," referred to as the reverse index sorting, that is, the number of documents containing a certain item in the document set. The higher the idf is, the lower its ability to distinguish the attributes of the document category, and the smaller the weight is.

(2)   KEA

On the basis of tf-idf, the position where words are first appeared in the article is added as a feature, mainly for the corpus of the total score structure. The words appearing in the head and tail of the article have a higher probability of becoming keywords, so different weights are given according to the position where each word appears in the first time. Based on this, tf-idf and continuous data discretization are used to extract keywords.

(3)   RAKE

Key phrases can be extracted. Candidates are generated from the stop list, and then, the scores of each candidate word and phrase are calculated. Finally, the final descending order is sorted based on score. When calculating the scores of each candidate word, the concept of word frequency and degree is used to generate a square matrix, which is also a diagonal matrix. The number on the diagonal line represents the word frequency of words, and the rest of column j in line i shows the frequency of the co-occurrence of word i and word j. For the word i, the degree is the sum of all the values in line i (theoretically a graph concept). Ultimately, the score of the word i refers to the value obtained (degree/frequency), while the score of the phrase refers to the sum of the scores of each word in the phrase.

(4)   TextRank

TextRank is an extension of PageRank. In this algorithm, each word in the text is regarded as a page, and a word in the text has a link with N words around it. Then, the weight of each word is calculated by using PageRank in this network, and the words with the highest weight are regarded as keywords. TextRank has some limitations, and the extracted words are also biased toward high-frequency words, because high-frequency words are more likely to form links with other words, and then, the corresponding words or phrases have a higher weight and ultimately form keywords.

(5) Algorithm based on left and right entropy

If a word is a key word, it will appear repeatedly, and the probability of different words from left to right is very high, that is, the left and right entropies are higher. In practice, this algorithm is mainly used for the discovery of new words. If it is used in the field of keyword extraction, the left and right entropies of all words are calculated. If the left and right information of a word is very large, then the word is likely to be a key word. If a word has a large information entropy on the left and a small information entropy on the right, and a small information entropy on the left and a large information entropy on the right, which can be described as X-B-C-Y, it can be considered that B and C often appear together. Combining B and C as key phrases, new words can be found based on the same principle. Similarly, the algorithm can be used for phrase discovery.

(6) Mutual information-based algorithms

Mutual information indicates the degree of interdependence between the two variables. The higher the value of mutual information, the greater the possibility of forming phrases. On the contrary, the lower the value is, the lower the possibility of word boundary between the two variables is. This algorithm is more suitable for phrase discovery.

6.1.3 Topic Detection and Tracking

Topic detection and tracking (TDT) originates from event detection and tracking (EDT), which is a new information processing technology proposed in recent years. This technology aims to help people deal with the increasingly serious Internet information explosion problem, to automatically identify new topics in news media information flow and to continuously track known topics. The object of TDT detection and tracking has expanded from events occurring at a specific time and place to topics with more relevant extensions. The corresponding theoretical and applied research has also leaped from the traditional recognition of events to topic detection and tracking including emergencies and their subsequent related reports.

The research started in 1996, and since 1998, an international TDT evaluation activity has been held every year. The evaluation was initiated by the Defense Department's Advanced Research Program (DARPA) and the National Bureau of Standards and Technology (NIST). Participants included government agencies such as DARPA, top universities such as CMU, Cambridge, and IBM, GE, and other companies. This evaluation has greatly promoted the development of TDT technology and achieved a lot of important research results.

Currently, the focus of TDT research is event detection and tracking. Theme is a broader concept than event, and a topic can contain multiple related events.

Essentially, event detection is clustering news streams according to different events. It is necessary to classify the reports discussing an event into one group. Compared with normal text clustering, the particularity of event detection is mainly manifested in two aspects: firstly, the object of event detection is the news report stream which appears in chronological order and changes with time, not a static closed text set; secondly, event detection is clustered according to the event discussed in the report rather than the subject category, and the information on which it is based. The granularity is relatively small, so more classes should be obtained from event detection.

In the process of event detection, the main steps are as follows: (1) To read a report from a data source, including content, time, and other relevant information; there may be multiple data sources, and there may be no clear boundaries between reports, which need to be preprocessed by segmentation between reports; (2) using centroid comparison or nearest neighbor comparison strategy to calculate the similarity between reports and events, or between reports and reports, to determine the events closest to current reports; (3) if a report is categorized as an event, then adjust the event; if the report cannot be categorized as an existing event, it will be categorized as a newly detected event; (4) output the detected events, using the most weighted feature words in the event, or a representative report title as the event description.

Theme detection and tracking processes the Internet as a large corpus according to different thematic strategies set by different users. The tracking system can link the post-inputted Internet reports with the topic by giving one or more reports on a topic. The actual implementation process is divided into two steps: given a set of sample information, the given topic model is obtained through training, and information related to the target topic is found in subsequent information. Because the object of topic recognition and tracking is a language information flow that changes with time, not a static and closed text set, the sequentiality of topic tracking is also considered.

The whole process is composed of three parts: subject model building, model-based tracking, and tracking result generating.

(1)   **Establishing theme model**

To judge whether the public opinion information is related to the theme, we need to solve the problem of the representation model of the theme. Here, we use vector space model to express the theme. The basic idea is: given a natural language document $D = D(t_1, w_1; t_2, w_2; \ldots; t_n, w_n)$, where t_i is the feature selected from document D, and w_i is the weight of the item, $1 \leq i \leq n$. Think of t_n as an n-dimensional coordinate system, while w_1, w_2, ..., w_n is the corresponding coordinate value, so $D(w_1, w_2, \ldots, w_n)$ is regarded as a vector in n-dimensional space. Documents are represented as vectors in document space, and words are used as text features. The weighting strategy of feature words is tf-idf. For each topic, it is necessary to build a known topic model through training information model. At this time, we need to calculate the similarity between information and topic, and use symmetrical Okapi formula algorithm to calculate the result, which is the score between document and topic.

(2) **Model tracking**

Tracking can be achieved through the relevant calculation process of keywords. By providing feedback manually, we can find out the message of the topic under discussion, get the weight λ_{i_n} of each keyword, and then calculate the score by selecting the scoring algorithm. In the topic set T, the relative probability of each keyword in is calculated as follows,

$$P(i_n|T) = \frac{tf(i, T)}{\sum_{j_n T} tf(j_n, T)}.$$

Relevant topic set T, irrelevant topic set T_0, and all topic set T_a are calculated using the above formula, where tf(i_n, X) represents the conditional probability of keyword in set X.

Next, by manually setting the weight λ_{i_n} of the keyword in, we can adjust the conditional probability of the keyword and standardize it,

$$P(i_n|T) = \lambda_i P(i_n|T) + (1 - \lambda_i)P(i_n|T_a),$$

$$P(i_n|T)_{\text{normal}} = \frac{tf(i, T)_{tz}}{\sum_{j_n T} tf(j_n, T)_{tz}}.$$

For the calculation in T_0, the same is done. Similar ratios #rol can be obtained by following calculations,

$$\#\text{rol} = \log \frac{P(i, T)_{\text{normal}}}{P(i, T_0)_{\text{normal}}}.$$

Finally, the score of message S in topic T is

$$\text{tfs}(i_n, S) = \text{tf(in, }S)\lambda_{i_n} / \sum_{j_n \in Ta} \text{tf}(j_n, T_a)\lambda_{j_n},$$

$$\text{score}(S, T) = 1 / \left(n_s \sum_{i_n \in d} \#\text{rol} * \text{Tfs(in, }S) \right).$$

In the formula, i_n is a key word that appears in message S, n_S is the number of different keywords in message S. Incremental learning mechanism is used to enhance the role of the current topic model, which is realized by merging new training information into the topic model.

(3) **Tracking results**

On the basis of the above algorithm, the tracking results are retracked through keyword adjustment and incremental learning mechanism, and the results are compared with the thresholds repeatedly to obtain more reliable tracking results.

At present, there is some room for improvement in the research of event detection at home and abroad, mainly reflected in: (1) Hot events ranking problem, people often do not have time to view a large number of news events, so the hottest news events ranking should be more advanced, which involves the design of ranking algorithm. (2) Event similarity problem, because the similarity of news reported on different aspects of the same news event may be small, so that the same news event is divided into several small events in the early stage of the event, and with the continuous development of the situation, the similarity of these events may become larger and larger, which may cause confusion and inconvenience to users' browsing. If a scientific similarity measurement method is given to fit the problem, the same content can be eliminated. (3) In the practical application environment, event detection is a long-term and continuous process. With the dynamic evolution of events, the relevance of some news in the event and the event is gradually decreasing. In addition, events with longer periods may also expand over time, and the content of the whole event is too broad. (4) There are two ways to describe news events at present: one is to select some of the most important feature words in the event, the other is to select a news headline in the event. Because Chinese natural language processing technology is not mature enough, it is often difficult to describe events accurately and completely. If a report title is used as a description, for some comprehensive events, the report may be only one aspect of the event, and the description of the event is not comprehensive enough.

At present, topic detection method is more effective than regular corpus such as news. The effect of irregular data such as forums (many short texts) and blogs (many very long texts) needs to be further improved. Special processing of irregular text such as short text and long text (including many contents) is needed to ensure better detection effect.

6.1.4 Automatic Summarization

With regard to the research of automatic abstract extraction, the first paper on automatic abstract published by H. P. Luhn of IBM Company in 1958 [6] opened the prelude of the research in this field. In recent years, a series of international conferences, such as Document Understanding Conference (DUC), ACM/IEEE Joint Digital Library Conference (JCDL), International Conference on Computational Linguistics (ACL), and International Information Retrieval Conference (SIGIR), all have the latest achievements on automatic summarization of documents. Several well-known systems in the field of automatic summarization of documents in the world, including the NEATS system of ISI, the NewsBlaster system of Columbia University, and the NewsIn Essence system of the University of Michigan, have greatly promoted the research in this field.

Automatic document summary can be divided into two categories according to the way it is generated: generative and extractive. Among them, extraction is to form abstracts by selecting the sentences in the original text. Usually, according to the predefined feature set, the sentences in the document are scored, and the output with high score is abstract sentences. Generative abstracts can contain words and phrases that do not appear in the original text, generally based on entity information, information fusion, and compression technology. Generally speaking, due to the relatively large difficulty of generative method, extraction method is widely studied at present.

(1)   Statistical-based extractive automatic abstraction process

In extractive automatic document summarization algorithm, a typical statistical-based algorithm step can be described as follows:

The first step is pretreatment, including coding conversion, sentence breaking, and so on.

The second step is feature extraction. Complete word segmentation, statistical word frequency, keyword extraction, named entity recognition, word (sentence) location information, paragraph structure information, etc.

The third step is to calculate the sentence weight. Calculate the weight of statements in a document.

The fourth step is to rank sentences according to their weights and extract abstract sentences according to their proportions.

The fifth step is to reorder the sentences according to the order in which they appear in the original text and do some polishing work such as anaphora resolution for the generated abstracts.

In the above steps, the most critical technology is the calculation of sentence weight in the third step. Generally speaking, the weight of a sentence is calculated by the weight of the words that make up the sentence. The weights of words are determined by some factors, such attention. The attention-based character-word hybrid neural network with semantic and structural information was proposed in [7]. In addition to calculating the weight of sentences by weighting various factors, we can also calculate the amount of information each sentence occupies for the full text from the perspective of information theory and select the sentences with the largest amount of information as abstract sentences. The information quantity of a sentence is the sum of the information quantity of each word in the sentence. The calculation of the information quantity of a word is based on the co-occurrence distribution of words and other words (statistical ngram).

(2)   Extractive automatic abstraction process based on machine learning

More and more researchers regard the task of automatic summarization as a binary classification problem at the sentence level and use machine learning to process it. The automatic summarization process based on machine learning can be summarized as five steps: feature selection, algorithm selection, model training, summarization extraction, and model evaluation.

The first step is feature extraction. It plays an important role in automatic abstracting. With the above statistical methods, typical features include named entity, proximity, entity word association, word location, and so on.

The second step is algorithm selection. Automatic summarization is a typical problem of sentence classification. It can be used in automatic summarization of sentence classification machine learning algorithms, including naive Bayesian, maximum entropy model, Bayesian network, support vector machine, and so on. Researchers choose the best classification algorithm according to the actual problem.

The third step is model training. The corpus is divided into training set and testing set. After training the algorithm with the training set, the performance of the algorithm is evaluated on the test set.

The fourth step is abstract extraction. On the issue of binary classification at sentence level, abstract sentences are generally defined as positive ones and non-abstract sentences as negative ones.

The fifth step is model evaluation. The main evaluation indices of automatic summarization based on machine learning include F value and ROUGE.

The recall-oriented understudy for gisting evaluation (ROUGE) evaluation method was proposed by Lin [8] after referring to the machine translation automatic evaluation method of Papineni. This method firstly generates artificial abstracts by several experts to form a standard abstract set and then compares the automatic abstracts generated by the system with the standard abstracts generated by man, and evaluates the quality of the abstracts by counting the number of basic units (n-ary grammar, word sequence, or word pairs) overlapped between them. The stability and robustness of the evaluation system are improved by comparing with the multi-expert artificial abstracts. ROUGE mainly includes the following four evaluation indicators:

1. ROUGE-N, based on ngram co-occurrence statistics,
2. ROUGE-L, based on the longest common substring,
3. ROUGE-S, based on sequential word pair statistics, and.
4. ROUGE-W, on the basis of ROUGE-L, considers the continuous matching of strings.

It is worth mentioning that in recent years, researchers have proposed a generative automatic summarization algorithm based on Seq2Seq model [9], which has achieved very good results. Overall, the model is based on an encoder–decoder structure, which first transforms the source sentence into a vector of a fixed dimension d and then generates a target sentence by a character in the decoder part. After adding attention distribution mechanism, decoder can get hidden state of the hidden layer of each character in the previous encoder encoding stage when generating a new target sequence, which improves the accuracy of generating a new sequence.

6.1.5 Sentiment Analysis

Text sentiment analysis refers to the process of analyzing, processing, and classifying subjective texts with emotional color by using natural language processing and text mining technology. The process of text affective analysis includes the whole process of crawling from the original text, text preprocessing, corpus, and affective lexicon construction, and affective analysis. The task of emotional analysis includes emotional classification, emotional retrieval, and emotional extraction. Among them, emotional classification is the most important task of emotional analysis. We elaborate on it here.

Emotional classification, also known as emotional orientation analysis, refers to the recognition of the positive or negative tendency of the subjective text for a given text. At present, the research work of text sentiment orientation analysis is mainly divided into two categories: semantic affective dictionary-based method and machine learning-based method.

To be specific, the rough process of identifying the text content as "positive" or "negative" mainly involves three steps, which is illustrated below:

Step 1: Crawl post-contents from Web page.

To crawl online text contents, we could use some crawler freeware, such as WebSiteSniffer and HTtrack. Usually, these tools could automatically capture all Web site files that are downloaded by the browser and save them to a local directory for offline browsing. However, it needs a further manual text selection if we conduct the sentiment analysis of post-content.

Additionally, it is more convenient to program web crawler by ourselves using Python or Java. In this way, a rudimentary knowledge of Cascading Style Sheets (CSS) in HTML is required. The web crawler starts with the specific URL to visit, parses the Web page source code, and crawls the post-contents by locating its specific CSS labels. Besides, some social networks provide API for researchers to crawl comment or post-contents from their Web sites, we could also take use of these APIs.

Step 2: Tokenize each post-text and get word bags.

Natural language toolkit (NLTK) is a suite of libraries and programs for symbolic and statistical natural language processing (NLP) for English written in the Python programming language. In this step, NLTK would be used to process the post-text.

Firstly, import the post-text as the corpus of NLTK; Secondly, tokenize the texts using the "nltk.word_tokenize()" method in NLTK, so that we could get each word of the sentence. Furthermore, prepositions (on, under, upon), adverbs (really, already, still), and conjunctions (and, but) make little contribution to the sentiment analysis of the texts, and hence, these words would be removed. In addition, users may use other languages like Korean and Arabic or some self-create words, these would also be removed. Ultimately, word stem processing would make words with different tenses and singular and plural forms into original words. To conclude, after this step, we get the selected word bag of each post-text.

Step 3: Make sentiment analysis using one of the three following methods.

Method 1: WordNet.

Introduction: WordNet is a lexical database for the English language. WordNet can be seen as a combination of dictionary and thesaurus, it gives a "+" or "−" score for every word. The database and software tools are freely available for download.

Based on WordNet, SentiWordNet[1] is a lexical resource in which each synset of WordNet is associated to three numerical scores Obj(s), Pos(s), and Neg(s), describing how objective, positive, and negative the terms contained in the synset are. Each of the three scores ranges from 0.0 to 1.0, and their sum is 1.0 for each synset.

Method description: In the easiest way, for each post-text, we could mark each word in its word bag a sentiment score according to WordNet or SentiWordNet and add up all words' sentiment scores, so that we could get the post-text's sentiment polarity as below. Besides, there are numerous research achievements pertinent to sentiment analysis using dictionary and corpus [11–13].

Aggregate score	Post-text polarity
Score > 0	Positive
Score = 0	Objective
Score < 0	Negative

Method 2: Support vector machine.

Introduction: Machine-learning approach relies on the famous machine learning algorithms to solve the sentiment analysis as a regular text classification problem that makes use of syntactic and linguistic features. Most popular machine learning algorithms include naive Bayes classifier (NB), neural network, SVM, deep learning, etc. Here, we would take SVM as an instance below [14, 15].

Method description: After the preprocessed data sets are collected in Steps 1 and 2, we train and deploy an SVM as a sentiment polarity classifier. There are three tasks when using an SVM for classification. First, the features of the post-contents have to be selected. Second, a data set used for training has to be labeled with its true classes. Third, the best combination of parameters and model setting has to be found [16, 17]. This step could be implemented by MATLAB, Java, or Python programming.

- Data set prepared in Steps 1 and 2 should be divided into TrainingSet and PredictionSet, and each of them should be conducted with a feature selection process.
- Concerning features selection, various syntactic and linguistic features of post-contents are available [18–23], which are given in Table 6.1.

However, none of these features have been widely accepted as the best feature selection method for sentiment classification or text categorization. Therefore, these features could be revised according to our empirical classification results. Additionally, there are numerous researches about features selection in machine learning concerning bag of words and syntactic structures, and we could develop a further research if it is needed.

- Then, FeatureVectors of TrainingSet should be used to train the raw SVM classifier. In this part, we could adopt the method (for instance "grid search") to find

---

[1] https://sentiwordnet.isti.cnr.it/.

Table 6.1 Features selection in machine learning

Feature	Description
Document frequency	Document frequency is the number of documents in which a term occurs
Mutual information	Mutual information score is a measure of the mutual dependence between the two classes
Information gain	Information gain ranks terms by considering their presence and absence in each class. A high score is assigned to terms that occur frequently in a class
Chi-square statistic	The chi score is a measurement related to both the frequency measurement and ratio measurement of two classes
Binormal separation	A score is set as a measurement of items related to different classes

out the optimal combination of parameters for the SVM classifier for optimized classification performance.

- As a result, we get a trained SVM classifier particularly fit for pro-vaccination or anti-vaccination classification problem.
- Finally, we could use this trained SVM classifier to predict the PredictionSet and other data that we need to process.

Method 3: Word vector similarity

Introduction:

In supervised learning methods, a large number of labeled training documents are required to train the learning machine. However, as data volume increased exponentially, it becomes difficult to manually create these numerous labeled training documents to meet training requirements. Therefore, many researches are working on semi-supervised or unsupervised learning methods which overcome these difficulties [16, 17].

To be specific, I have proposed "A semi-supervised sentiment analysis algorithm based on similarity of word embedding" that in a project I participated, the details of this algorithm would be shown below.

Preparation:

Before starting the analysis process, we need to prepare three dictionaries:

· *Dictionary1*: *SeedWords.txt.*

The SeedWords.txt is the document of the seed dictionary in sentiment analysis, which is manually collected for this algorithm. It contains M positive seed words and N negative seed words particularly adapting to comments in vaccination field.

· *Dictionray2*: *SynonymDic.txt.*

The SynonymDic.txt is an existing synonym dictionary which can be searched online.

· *Dictionray3*: *PosSeedVectors.bin and NegSeedVectors.bin.*

The PosSeedVectors.bin and NegSeedVectors.bin are the word embedding documents of the positive seed words and the negative seed words, respectively. In these two documents, the positive and negative seed words are represented by vectors in the approach of distributed representation.

Procedure Description:

We adopt the word bag of each post-text accomplished in step two, and extract each post-word in word bag to make a sentiment analysis through the following procedure:

1. We make a comparison with the seed dictionary "SeedWords." If the post-word is included in "SeedWords," its sentiment polarity will be designated in accordance to "SeedWords." Otherwise, take the post-word into the next step.
2. We extract all the synonyms of the post-word, and further compare them with "SeedWords." As far as we find a synonym is included in "SeedWords," the sentiment polarity of this post-word would be consistent with the synonym in "SeedWords." Otherwise, the post-word will enter into the next step.
3. We use the distributed representation method to represent the term vector of this post-word as $w = (w_1, w_2, \ldots, w_j)$, the word2vec tool developed by Google will be implemented during this process.
4. We take out the previous PosSeedVectors.bin, which is defined as

$$
\vec{P} = \begin{bmatrix} p_{11} & p_{12} & \cdots & p_{1j} \\ p_{21} & p_{22} & & p_{2j} \\ \cdots & & \cdots & \cdots \\ p_{M1} & p_{M2} & \cdots & p_{Mj} \end{bmatrix}.
$$

In this matrix, each row means a seed word's vector, and there are M positive words' vectors. Furthermore, we calculate the similarity value between each seed word's vector (each row) and the post-word and define the result as SP_i. The calculation formula is shown as below,

$$
SP_i = \cos(\theta) = \frac{w \cdot p_i}{\|w\| \|p_i\|} = \frac{\sum_{k=1}^{j} w_k \times p_k}{\sqrt{\sum_{k=1}^{j} (w_k)^2} \times \sqrt{\sum_{k=1}^{j} (p_k)^2}}.
$$

Finally, we add up all the similarity value of the MSP_i results and get the similarity value between the post-word and positive seed words, defined as SP,

$$
SP = \sum_{i=1}^{M} SP_i.
$$

5. In the same way, we take out the previous NegSeedVectors.bin, which is defined as

$$
\vec{N} = \begin{bmatrix} n_{11} & n_{12} & \cdots & n_{1j} \\ n_{21} & n_{22} & & n_{2j} \\ \cdots & & \cdots & \cdots \\ n_{N1} & n_{N2} & \cdots & n_{Nj} \end{bmatrix}.
$$

In this matrix, each row means a seed word's vector, and there are N negative words' vectors. Furthermore, we calculate the similarity value between each seed word's vector (each row) and the post-word and define the result as SN_i. The calculation formula is shown as below,

$$SN_i = \cos(\theta) = \frac{w \cdot n_i}{\|w\|\|n_i\|} = \frac{\sum_{k=1}^{j} w_k \times n_k}{\sqrt{\sum_{k=1}^{j}(w_k)^2} \times \sqrt{\sum_{k=1}^{j}(n_k)^2}}$$

Finally, we add up all the similarity value of the NSN_i results and get the similarity value between the post-word and negative seed words, defined as SN,

$$SN = \sum_{i=1}^{N} SN_i$$

6. We compare the value of SP and SN and designate the sentiment polarity of the post-word in keeping with the larger one,

$$\begin{cases} SP > SN, postword = positive, \\ SP = SN, ostword = neutral, \\ SP < SN, postword = negative. \end{cases}$$

7. Add up all post-words' sentiment scores, so that we could get the post-text's sentiment polarity as below,

Aggregate score	Post-text polarity
score > 0	positive
score = 0	objective
score < 0	negative

6.2 Online Social Networks Image Recognition Models

Computer vision includes the following branches: picture reconstruction, event monitoring, target tracking, target recognition, machine learning, index building, image restoration, etc.

6.2.1   Image Retrieval

Since the 1970s, the research on image retrieval has begun. At that time, it was mainly based on text-based image retrieval technology (TBIR), which used the way of text description to describe the characteristics of images, such as the author, age, genre, and size of paintings. After the 1990s, the content semantics of images, such as color, texture, layout, and so on, were analyzed and retrieved by image retrieval technology, namely content-based image retrieval (CBIR) technology. CBIR belongs to content-based retrieval (CBR). CBR also includes the retrieval technology of dynamic video, audio, and other forms of multimedia information.

Therefore, image retrieval can be divided into two categories according to different ways of describing images, one is text-based image retrieval, and the other is content-based image retrieval. Text-based image retrieval mainly uses text annotation to add keywords to images, such as objects and scenes of images. When retrieving an image, the desired image can be retrieved directly according to the keywords to be searched. This method is simple to implement, but it is very labor-intensive (each image needs to be labeled artificially), and it is not realistic for large-scale database retrieval. In addition, there are human cognitive errors in manual annotation. People have different understandings of the same image, which leads to inconsistent annotation, which makes text-based image retrieval gradually lose luster. Content-based image retrieval technology is based on the content features of the image itself to retrieve the image, which eliminates the process of labeling the image artificially. It uses some algorithm to extract the features of the image and stores the features to form the image feature database. When the image needs to be retrieved, the same feature extraction technology is used to extract the features of the image to be retrieved, and the correlation between the image and the image to be retrieved in the feature database is calculated according to some similarity criteria. Finally, the image most relevant to the image to be retrieved is obtained by sorting from large to small, and image retrieval is realized. This method makes the retrieval process automated. The quality of image retrieval results depends on the quality of image feature extraction. In the face of massive data retrieval environment, we also need to consider the process of image comparison (image similarity considerations). In practice, it is also very important to use efficient algorithms to quickly find similar images [24].

(1) **Image retrieval of same object**

Image retrieval of the same object refers to searching an object in the image to find the image containing the object from the image database. Here, the user is interested in the specific object or target contained in the image, and the retrieved image should contain those pictures of the object. For example, we present a "Mona Lisa" image. The goal of the same object retrieval is to retrieve images containing "Mona Lisa" characters from the image database. After sorting by similarity measure, these images containing "Mona Lisa" characters are ranked as far as possible in front of the retrieval results. Similar object retrieval is generally called object retrieval in English literature. Duplicate search or detection can also be classified as the same object retrieval, and the same object retrieval method can be directly applied to similar sample search or detection. For image retrieval of the same object, when retrieving the same object or target, it is vulnerable to the influence of the photographic environment, such as the change of illumination, scale, angle of view, occlusion, and background clutter. Due to the large environmental interference, for the same object image retrieval, when selecting features, some invariant local features with better anti-interference are often selected, such as scale-invariant feature transform (SIFT) [25], speeded up robust features (SURF) [26], oriented FAST and rotated BRIEF (ORB), etc. [27], and based on this, construct a global description of the image by different coding methods. Representational work includes word bag model (bag of words) [28], local feature aggregation descriptor (VLAD, vector of locally aggregated descriptors), and Fisher vector (FV). This kind of SIFT-based image retrieval method combines the invariant characteristics of SIFT-like and adopts the local to global feature expression. In practical application, siftGPU can also be used to accelerate the extraction of SIFT when extracting SIFT, so as to achieve better retrieval results on the whole. But this kind of method usually has very high feature dimension. In order to obtain high retrieval accuracy, the number of clustering is usually set to hundreds of thousands. Therefore, it is necessary to design an efficient indexing method for them.

(2) **Image retrieval of the same category**

For a given query image, the goal of similar image retrieval is to find out those images belonging to the same category as the given query image from the image database. Here, the user is interested in the category of objects and scenes, that is, the user wants to obtain the images of objects or scenes with the same category attributes. In order to better distinguish between the same object retrieval and the same category retrieval, we still take "Mona Lisa" as an example. If the user is interested in the picture of "portrait" rather than the painting itself, that is to say, the user is interested in abstracting the concept of category of the specific painting, then the retrieval system should search in the same way. The same category of image retrieval has been widely used in image search engine, medical image retrieval, and other fields.

For image retrieval of the same category, the main problem is that the intra-class changes of images belonging to the same category are enormous, while the differences between different image classes are small [29, 30]. Therefore, the intra-class changes and the inter-class differences in feature description of image retrieval of the same category exist great challenges. In recent years, when automatic features based on deep learning (DL) are applied to image retrieval of the same category, they can greatly improve the accuracy of retrieval and make the same object-oriented retrieval get a better solution in feature expression. At present, the feature expressions dominated by convolutional neural network (CNN) are also beginning to be developed on the same object image retrieval, and some corresponding work has been done [31], but the same object is in the construction class. Sample training data is not as convenient as the same type of image retrieval, so the same object image retrieval needs to be further studied in CNN model training and automatic feature extraction. Whether it is the same object image retrieval or the same category image retrieval, when using CNN model to extract automatic features, the final dimension is generally a 4096-dimensional feature, and its dimension is relatively high. Direct use of PCA and other dimension reduction means can achieve the purpose of feature dimension reduction. However, the dimensionality that can be reduced is still limited while maintaining the necessary retrieval accuracy. Therefore, it is also necessary to construct efficient and reasonable fast retrieval mechanism for this kind of image retrieval, so that it can adapt to large-scale or massive image retrieval.

(3) **Features of large-scale image retrieval**

Whether for the same object image retrieval or the same category image retrieval, they have three typical main features on large-scale image data sets: large image data volume, high feature dimension, and short time required. The following three main features are explained one by one:

1. Large amount of image data. Thanks to the development of multimedia information acquisition, transmission, storage, and the improvement of computer operation speed, after more than ten years' development, the scope of image retrieval technology based on content has expanded from the original small image database to the large-scale image database or even the large-scale image database. For example, in the early stage of the development of image retrieval technology in the 1990s, researchers verified the performance of image retrieval algorithms. The image library has a total of 1000 images, which can be used for image retrieval as well today. Compared to the image net data set of the most popular image classification library, its magnitude has grown by tens of thousands of times. Therefore, image retrieval should meet the requirements of the era of big data and should be scalable on large-scale image data sets.
2. The feature dimension is high. As the cornerstone is directly describing the visual content of an image, image features directly determine the highest retrieval accuracy that can be achieved in the retrieval process. If the prefeature is not well expressed, the post-retrieval model will not only complicate the construction of the model, increase the response time of the retrieval query, but also improve the retrieval accuracy is extremely limited. So, at the beginning of feature extraction, we should consciously select those higher-level features. If local feature expression is also considered as a kind of "high dimension," then the descriptive ability of feature has a great correlation with the dimension of feature. So the large-scale image retrieval has obvious characteristics of high dimension in feature description, such as bag of words model, VLAD, Fisher vector, and CNN features.
3. Require fast response. For user queries, image retrieval system should have the ability to respond quickly to user queries. At the same time, because of the large amount of large-scale image data and high feature dimension, it is difficult to meet the real-time requirements of the system by directly using brute search index strategy (also known as linear scanning). Therefore, large-scale image retrieval needs to solve the problem of real-time response of the system.

(4) **Nearest Neighbor Search Algorithms**

In recent years, in order to reduce the space complexity and time complexity of the search space, the researchers have found an approximate nearest neighbor (ANN) search method and proposed many efficient search techniques, among which the most successful. The method includes an image retrieval method based on a tree structure, a hash-based image retrieval method, and an image retrieval method based on vector quantization.

Tree-based image retrieval method organizes the corresponding features of the image with tree structure, which reduces the computational complexity to the logarithmic complexity of the sample number of the image database. The search methods based on tree structure include KD-tree, M-tree, and so on. Among many tree structure search methods, KD-tree is the most widely used one. A simple KD-tree partitioning process [32]: In the search stage, after the query data from the root node to the leaf node, the data under the leaf node are compared with the query data one by one, and the way of backtracking is used to find the nearest neighbor. Although tree-based retrieval technology greatly reduces the response time of a single search, the performance of tree-based indexing method will decline sharply when the dimension of high-dimensional features is several hundred, or even close to or lower than that of violent search. In addition, tree-based retrieval methods tend to occupy much more storage space than the original data when constructing tree structure and are sensitive to data distribution, which makes tree-based retrieval methods face the problem of memory constraints in large-scale image databases [33].

The hash-based image retrieval method can encode the original feature into a compact binary hash code, so that the hash-based image retrieval method can greatly reduce the memory consumption and can be used when calculating the Hamming distance. The internal arithmetic of the computer has an XOR operation, so that the Hamming distance calculation can be completed in the order of microseconds, which greatly reduces the time required for a single query response. The key point of the algorithm is to design an effective hash function set, so that the data in the original space can be mapped by the hash function set, and the similarity between the data in Hamming space can be better maintained or enhanced. The most classical hashing method is locality-sensitive hashing (LSH). It uses the method of random hyperplane to construct hash function. Even if the space is divided into many sub-regions by random hyperplane, each sub-region can be regarded as a "bucket." In the construction stage, local-sensitive hashing only needs to generate random hyperplanes, so there is no training process; in the indexing stage, samples are mapped to binary hash codes, and samples with the same binary hash codes are stored in the same "bucket"; in the query stage, the query sample can be locked in which "bucket" after the same mapping, and then in the locked "bucket," the query sample and the sample in the "bucket" are compared one by one, so as to get the final nearest neighbor. The validity of locally sensitive hashing is guaranteed in theoretical analysis, but because the data itself is not used in the construction of the hash function by locally sensitive hashing, long coding bits are often used in the application of locally sensitive hashing to obtain higher spermatic accuracy, but the collision probability of similar samples in the process of hashing discretization is reduced under the condition of long coding bits. As a result, the recall rate of retrieval will decrease greatly.

A typical representative of vector quantization-based methods is the product quantization (PQ) method, which decomposes the feature space into Cartesian products of multiple low-dimensional subspaces and then quantizes each subspace separately. In the training phase, each subspace is clustered to obtain a centroid (i.e., quantizer), and the Cartesian product of all these centroids constitutes a dense partition of the whole space and can ensure that the quantization error is relatively small; after learning, for a given query sample, the asymmetric distance of the sample in the query and the library can be calculated by looking up in table [34]. Although the product quantization method is more accurate when approximating the distance between samples, the data structure of the product quantization method is usually more complicated than the binary hash code, and it cannot obtain low-dimensional feature representation, and in order to achieve good performance, An asymmetrical distance must be added, and it also requires a balance of variance for each dimension. If the variance is not balanced, the product quantization method yields poor results.

6.2.2 Image Object Detection and Classification

Image object detection and classification is a very active research direction in the fields of computer vision, pattern recognition, and machine learning, and has been widely used in many fields, including face recognition in security field, pedestrian detection, intelligent video analysis, pedestrian tracking, traffic scene object recognition, vehicle counting, retrograde detection, license plate detection and recognition, and content-based image retrieval in the Internet field, automatic album categorization, etc. Image object detection and classification tasks require answering whether an image contains an object. Characterizing the image is the main research content of object classification [35].

Generally speaking, object classification algorithm describes the whole image globally by manual feature or feature learning method and then uses classifier to determine whether there is a class of objects. Object detection task is more complex. It needs to answer what object exists in an image at what position. Therefore, besides feature expression, object structure is the most important difference between object detection tasks and object classification. Generally speaking, in recent years, object classification methods mostly focus on learning feature expression. Typical methods include bag of words model and deep learning model. Object detection methods focus on structure learning, which is represented by deformable component model.

(1) Object classification based on word packet model

Bag of words is the basic framework of object classification algorithm. It originally came into being in the field of natural language processing. It describes and expresses documents by modeling the frequency of words appearing in documents. A lot of research work has focused on the study of word packages model, and gradually formed a standard target classification framework consisting of four parts: bottom features, feature coding, feature aggregation, and classifier design.

1. Bottom feature extraction. Bottom feature is the first step in the framework of object classification and detection. There are two methods of bottom feature extraction: one is based on interest point detection; the other is based on intensive extraction. Interest point detection algorithm chooses clearly defined pixels, edges, corners, blocks with obvious local texture features by some criteria, and usually obtains certain geometric invariance, so that it can get more meaningful expression at a smaller cost. The most commonly used interest point detection operators are Harris corner detector and features from accelerated segment test (FAST) operator, Laplacian of Gaussian (LoG), difference of Gaussian (DoG), etc. In recent years, more and more intensive extraction methods have been used in object classification field. A large number of local feature descriptions are extracted from images according to fixed step size and scale. Although a large number of local descriptions have higher redundancy, they have more abundant information. After effectively expressing with the word package model, we can usually get better performance than interest point detection. Common local features include scale-invariant feature transform (SIFT), histogram of oriented gradient, directional gradient histogram (HOG) [36], and local binary pattern, local binary pattern (LBP) [37]. Many features are used in the object classification algorithm with good performance. Intensive extraction in sampling mode is combined with interest point detection. Many feature descriptors are also used in the bottom feature description. The advantage of this method is that in the bottom feature extraction stage, a large number of redundant features are extracted to maximize the bottom description of the image and to prevent the loss of too much useful information. Redundant information in some underlying descriptions is abstracted and simplified mainly by feature coding and feature aggregation [38].
2. Feature coding. The underlying features of dense extraction contain a lot of redundancy and noise. To improve the robustness of feature expression, a feature transformation algorithm is needed to encode the underlying features to obtain more distinguishing and more robust. In the characteristic expression, this step plays a crucial role in the performance of object recognition. Important feature coding algorithms include vector quantization coding, kernel dictionary coding [39], sparse coding [40], local linear constraint coding [41], saliency coding [42], Fisher vector coding [43], super vector coding [44], etc.
3. Vector quantization coding. It uses a small feature set (visual dictionary) to describe the underlying features and achieves the purpose of feature compression. Soft quantization coding (also known as kernel visual dictionary coding) algorithm's local features are no longer described by a visual word but are described by weighted K visual words nearest to each other, which effectively solves the ambiguity of visual words and improves the accuracy of object recognition. Sparse coding achieves sparsity of response on an over-complete basis by adding

sparse constraints to least squares reconstruction. Local linear constrained coding reconstructs the underlying features on a local manifold by adding local linear constraints, which not only ensures that the resulting feature coding does not have the discontinuity of sparse coding, but also preserves the sparse feature of sparse coding. Significance coding introduces the concept of visual saliency. If the distance between a local feature and the nearest and the next nearest visual word is very small, it is considered that the local feature is not "significant" and the response after coding is also small. Supervector coding and Fisher vector coding are the best feature coding methods proposed in recent years. Their basic ideas are similar, which can be considered as the difference between coding local features and visual words. Fisher vector coding combines the abilities of both production model and discriminant model. Unlike traditional feature coding based on reconstruction, it records the first- and second-order differences between local features and visual words. Supervector coding directly uses the difference between local features and the most recent visual words to replace the previous simple hard voting.

4. Feature aggregation. Spatial feature aggregation is a feature set integration operation after feature coding. By taking the maximum or average value of each dimension of the encoding feature, a compact feature vector is obtained as the feature expression of the image. At the same time, it avoids the high cost of using feature set for image expression.

5. Classification using support vector machines and other classifiers. From image extraction to feature expression, an image can be described by a fixed dimension vector. Next, we learn a classifier to classify the image. At this time, there are many classifiers that can be selected. The commonly used classifiers are support vector machine, K-nearest neighbor, neural network, random forest, and so on. As one of the most widely used classifiers, SVM based on maximizing boundaries performs well in image classification tasks, especially in SVM using kernel method.

(2) **Deep learning model**

Deep learning model is another kind of object recognition algorithm [45]. Its basic idea is to learn hierarchical feature expression in supervised or unsupervised way to describe objects from the bottom to the top. Mainstream deep learning models include auto-encoder [46], restricted Boltzmann machine (RBM) [47], deep belief nets (DBN) [48], convolutional neural network [49], and bioheuristic model.

Auto-encoder is a special neural network structure proposed in the 1980s. It has been widely used in data dimensionality reduction and feature extraction. The automatic encoder consists of an encoder and a decoder. The encoder transforms the data input into the hiding layer representation, and the decoder is responsible for recovering the original input from the hiding layer. The number of hidden layer units is usually less than the dimension of data input, which acts as a "bottleneck" and keeps the most important information in the data, thus realizing data dimension reduction and feature coding. Automatic encoder is an unsupervised feature learning unit based on feature reconstruction. Different constraints can be added to it, and different changes can be obtained, including denoising auto-encoders and sparse auto-encoders. Good results have been achieved in digital handwriting recognition and image classification tasks.

Restricted Boltzmann machine is an undirected bipartite graph model and a typical energy-based model (EBM). The term "constrained" refers to the connection between the visual layer and the hidden layer, but there is no connection between the visual layer and the hidden layer. The special structure of the restricted Boltzmann machine makes it conditionally independent, that is, given the hidden layer units, the visual layer units are independent and vice versa. This feature enables parallel Gibbs sampling of cells in one layer at the same time. Contrastive divergence (CD) algorithm is usually used for model learning of restricted Boltzmann machines.

Deep belief network (DBN) is a hierarchical undirected graph model. The basic unit of DBN is restricted Boltzmann machine (RBM). First, the original input is used as the visual layer to train a single layer of RBM, then the weight of the first layer of RBM is fixed, and the response of RBM hidden layer unit is used as the new visual layer to train the next layer of RBM, and so on. Through this kind of greedy unsupervised training, the whole DBN model can get a better initial value, and then, tag information can be added. The whole network can be supervised and fine tuned by means of production or discriminant to further improve network performance. DBN has been successfully applied in handwritten digital recognition, speech recognition, content-based retrieval, and other fields.

Convolutional neural network was first used in digital handwriting recognition in the 1980s and achieved some success. However, due to hardware constraints, the high computational cost of convolutional neural network makes it difficult to apply it to actual size target recognition tasks. Convolutional neural network mainly includes convolution layer and convergence layer. Convolution layer simulates simple cells by convoluting the whole image with a fixed size filter. Convergence layer is a kind of down-sampling operation, which achieves the goal of down-sampling by taking the maximum and average values of local blocks in the convolution feature map and obtains some invariance in the process. Convergence layers are used to simulate complex cells in theory. After the response of each layer, there are usually several nonlinear transformations, such as sigmoid, tanh, relu, and so on, which enhance the expressive ability of the whole network. At the end of the network, a number of fully connected layers and a classifier are usually added, such as soft Max classifier, RBF classifier, and so on. The filter of convolution layer in convolution neural network is shared in all locations, so it can greatly reduce the scale of parameters, which is very useful to prevent the model from being too complex [50]. On the other hand, the convolution operation keeps the spatial information of the image, so it is especially suitable for image expression.

Here, we compare the most popular word package model with the convolution neural network model and find that they are very similar. In the word package model, the process of feature coding for the underlying features is approximately equivalent to the convolution layer in the convolution neural network, and the operation of the convergence layer is the same as the convergence operation in the word package model. The difference is that the word package model actually contains only one convolution layer and one convergence layer, and the model uses unsupervised learning method for feature expression. The convolution neural network contains more layers of simple and complex cells, which can carry out more complex feature transformation, and its learning process is supervised. The filter weight can be based on data and tasks. Make constant adjustments to learn more meaningful feature expression. From this point of view, convolutional neural network has more powerful feature expression ability, and its excellent performance in image recognition task can be easily explained.

(3) **Object detection**

The most important difference between object detection task and object classification task is that object structure information plays an important role in object detection, while object classification is more concerned with the global representation of objects or images. The input of object detection is a window that contains objects, while object classification is the whole image. For a given window, object classification and object detection are very similar in feature extraction, feature coding, and classifier design.

According to the different strategies of obtaining window position, object detection methods can be roughly divided into sliding window and generalized Hough transform. The sliding window method is relatively simple. It uses the trained template to scan the input image on multiple scales and finds the external window of the target object by determining the maximum response position. Sliding window method has been widely used in practice because of its simplicity and effectiveness. Especially, the appearance and development of histograms of oriented gradients (HOG)) model and deformed component model make sliding window model become the mainstream object detection method. Generalized Hough's voting method is a method to get the position of an object by accumulating in the parameter space. It can be used for arbitrary shape detection and general object detection tasks.

Different from object classification, object detection studies the relationship between input image X and output object window Y mathematically. Here, the value of Y is no longer a real number, but a group of "structured" data, which specifies the external windows and categories of objects. It is a typical structured learning problem. Structured SVM (SSVM) [51] based on maximizing edge criterion generalizes ordinary support vector machine to handle structured output, effectively expands the scope of application of support vector machine, can handle more general data structures such as grammar tree, graph, etc., and receives more and more attention in natural language processing, machine learning, pattern recognition, computer vision, and other fields [52]. Note that, latent SVM (LSVM) is used to deal with the problem of object detection. The basic idea of LSVM is to optimize the object position as a hidden variable into the objective function of SVM and to obtain the optimal object position by discriminant method. Weak-label structural SVM (WL-SSVM) is a more general framework for structured learning. Its main purpose is to deal with the problem of inconsistency between label space and output space. When multiple outputs conform to one label, each sample label is considered as a "weak label.". Both SSVM and LSVM can be regarded as special cases of WL-SSVM.

WL-SSVM can be transformed into general SSVM and LSVM through some reduction. Conditional random field (CRF), as a classical structured learning algorithm, has attracted considerable attention in object detection tasks [53]. By modeling the label of object components as hidden nodes and using EM algorithm to learn, the algorithm overcomes the shortcomings of traditional CRFs that need to manually give a topological structure, can automatically learn more flexible structures, and automatically discover meaningful component expressions in visual semantics in complicated environments.

References

1. Augenstein I, Derczynski L, Bontcheva K (2017) Generalisation in named entity recognition: a quantitative analysis. Comput Speech Lang 44:61–83
2. Silberer C, Frank A (2017) Casting implicit role linking as an anaphora resolution task. In: Joint conference on lexical and computational semantics. Association for Computational Linguistics, pp 1–10
3. Charniak E, Elsner M (2009) EM works for pronoun anaphora resolution. Conf of the European Chapter of the Association for Computational Linguistics. Association for Computational Linguistics, pp 148–156
4. Surdeanu M, Tibshirani J, Nallapati R (2012) Multi-instance multi-label learning for relation extraction. Joint conference on empirical methods in natural language processing and computational natural language learning, pp 455–465.
5. Datta S, Das S (2019) Multiobjective support vector machines: handling class imbalance with Pareto optimality. IEEE Trans Neural Netw Learn Syst 30(5):1602–1608
6. Luhn HP (1958) The automatic creation of literature abstracts. IBM Corp.
7. Guo SX, Sun X, Wang S, Gao Y, Feng J (2019) Attention-based character-word hybrid neural networks with semantic and structural information for identifying of urgent posts in mooc discussion forums. IEEE Access 7:120522–120532
8. Lin CY (2004) A package for automatic evaluation of summaries. The Workshop on Text Summarization Branches Out, p 10
9. Nallapati R, Zhou B, Santos CND (2016) Abstractive text summarization using sequence-to-sequence RNNs and beyond. In: SIGNLL conference on computational natural language learning
10. Medhat W, Hassan A, Korashy H (2014) Sentiment analysis algorithms and applications: a survey. Ain Shams Eng. J. 5(4):1093–1113
11. Maite T, Julian B, Milan T (2011) Lexicon-based methods for sentiment analysis. Association for Computational Linguistics
12. Akshi K, Teeja MS (2017) Sentiment analysis on twitter. Int. J. Comput. Sci. Issues 9(4)
13. Janyce W, Rebecca B (2001) A corpus study of evaluative and speculative language. In: Proceedings of the 2nd ACL SIG on Dialogue Workshop on Discourse and Dialogue (Aalborg, Denmark)
14. Li Y, Li T (2013) Deriving market intelligence from microblogs. Decis Support Syst 55(1):206–217
15. Chen C, Tseng Y (2011) Quality evaluation of product reviews using an information quality framework. Decis Support Syst 50(4):755–768
16. Xu W, Tan Y (2020) Semisupervised text classification by variational autoencoder. IEEE Trans. Neural Netw. Learn. Syst. 31(1):295–308
17. Zhang R, Nie F, Li X (2019) Semisupervised learning with parameter-free similarity of label and side information. IEEE Trans. Neural Netw. Learn. Syst. 30(2):405–414

18. Xue B (2011) Predicting consumer sentiments from online text. Decis Support Syst 50(4):732–742
19. Li S, Xia R, Zong C (2009) A framework of feature selection methods for text categorization. In: Joint Conference of the, Meeting of the ACL, pp 692–700
20. Berry MW, Kogan J (2010) Text mining: applications and theory (2010)
21. Ren K, Yang H, Zhao Y, Chen W, Xue M, Miao H, Huang S, Liu J (2019) A robust auc maximization framework with simultaneous outlier detection and feature selection for positive-unlabeled classification. IEEE Trans. Neural Netw Learn Syst 30(10):3072–3083
22. Forman G (2003) An extensive empirical study of feature selection metrics for text classification. J Mach Learn Res 3(2):1289–1305
23. Andrew KM, Sebastian T (2000) Text classification from labeled and unlabeled documents using EM. Mach Learn 39(2–3):103–134
24. Lu X, Zheng X, Li X (2017) Latent semantic minimal hashing for image retrieval. IEEE Press
25. Shen H, Liang X, Wang M (2016) Emergency decision support architectures for bus hijacking based on massive image anomaly detection in social networks. In: IEEE international conference on systems, man, and cybernetics, pp 864–869
26. Bay H, Tuytelaars T, Gool L J. Surf: speeded up robust features. Proc of IEEE Int. Conf. Comput. Vis., 404–417, 2006.
27. Rublee E, Rabaud V, Konolige K (2011) ORB: an efficient alternative to SIFT or SURF. In: Proceedings of IEEE international conference on computer vision, pp 2564–2571
28. Csurka G, Dance C, Fan L (2004) Visual categorization with bags of key points, workshop on statistical learning in computer vision. Eur. Conf. Comput. Vis. 1:1–2
29. Rossi RA, Zhou R (2020) Ahmed N K, Deep inductive graph representation learning. IEEE Trans Knowl Data Eng 3:438–452
30. He F, Nie F, Wang R, Li X, Jia W (2020) Fast semisupervised learning with bipartite graph for large-scale data. IEEE Trans Neural Netw Learn Syst 31(1):626–638
31. Kiapour MH, Han X, Lazebnik S (2015) Where to buy it: matching street clothing photos in online shops. In: Proceedings of IEEE international conference on computer vision, pp 3343–3351
32. Bentley JL (1975) Multidimensional binary search trees used for associative searching. Commun ACM Conf 18(9):509–517
33. Zheng C, Wang S, Liu Y, Liu C, Xie W, Fang C, Liu S (2019) a Novel Equivalent Model of Active Distribution Networks Based on LSTM. 30(7):2611–2624
34. Dong W, Charikar M, Li K (2008) Asymmetric distance estimation with sketches for similarity search in high-dimensional spaces. In: Proceedings of ACM SIGIR conference of research and development in information retrieval, pp 123–130
35. Zheng Y, Sun L, Wang S, Zhang J, Ning J (2019) Spatially regularized structural support vector machine for robust visual tracking. IEEE Trans Neural Netw Learn Syst 30(10):3024–3034
36. Dalal N, Triggs B (2005) Histograms of oriented gradients for human detection. In: Proceedings of the computer vision and pattern recognition, San Diego, CA, USA, pp 886–893
37. Timo O, Matti P, Topi M (2002) Multiresolution gray-scale and rotation invariant texture classification with local binary patterns. IEEE Trans Pattern Anal Mach Intell 24:971–987
38. Song J, Guo Y, Gao L, Li X, Hanjalic A, Shen H (2019) From deterministic to generative: multimodal stochastic RNNs for video captioning. IEEE Trans Neural Netw Learn Syst 30(10):3047–3058
39. Jan C, Jan M, Cor J, Arnold W (2008) Kernel codebooks for scene categorization. In: Proceedings of the European conference on computer vision, Marseille, France, pp 696–709
40. Bruno A, David J (1997) Sparse coding with an overcomplete basis set: a strategy employed by v1. Vision Res 37:3311–3325
41. Wang J, Yang J, Yu K, Lv F, Huang T, Gong Y (2010) Locality-constrained linear coding for image classification. In: Proceedings of the IEEE conference on computer vision and pattern recognition, San Francisco, CA, USA, pp 3360–3367
42. Huang Y, Huang K, Yu Y, Tan T (2011) Salient coding for image classification. In: Proceedings of the IEEE conference on computer vision and pattern recognition, Colorado, USA, pp 1753–1760

43. Florent P, Jorge S, Thomas M (2010) Improving the fisher kernel for large-scale image classification. In: Proceedings of the European conference on computer vision, Crete, Greece, vol 6314, pp 143–156

44. Zhou X, Yu K, Zhang T, Thomas S (2010) Image classification using super-vector coding of local image descriptors. In: Proceedings of the European conference on computer vision, Berlin, Heidelberg, pp 141–154

45. Bengio Y (2009) Learning deep architectures for AI. Foundations and Trends in machine learning. ISSN 1935-8237, Now Publisher Inc.

46. Bourlard H, Kamp Y (1988) Auto-association by multilayer perceptrons and singular value decomposition. Biol Cybern 59:291–294

47. Smolensky P (1986) Information processing in dynamical systems: foundations of harmony theory. In: Processing of the parallel distributed: explorations in the microstructure of cognition, vol 1: Foundations. MIT Press

48. Hinton GE, Osindero S, Teh Y (2006) A fast learning algorithm for deep belief nets. Neural Comput 18(7):1527–1554

49. LeCun Y, Bottou L, Bengio Y, Haffner P (1998) Gradient-based learning applied to document recognition. Proc IEEE 86(11):2278–2324

50. Deng X, Yan C, Ma Y (2020) PCNN mechanism and its parameter settings. IEEE Trans Neural Netw Learn Syst 31(1):488–501

51. Tsochantaridis B, Joachims T, Hofmann T, Altun Y (2006) Large margin methods for structured and interdependent output variables. J Mach Learn Res 6(2):1453–1484

52. Zheng J, Cao X, Zhang B, Zhen X, Su X (2019) Deep ensemble machine for video classification. IEEE Trans Neural Netw Learn Syst 30(2):553–565

53. Jin Y, Xie J, Guo W, Luo C, Wu D, Wang R (2019) LSTM-CRF neural network with gated self attention for Chinese NER. IEEE Access 7:136694–136703

Chapter 7
Social Networks Structure Analysis and Online Community Discovery

7.1 Online Social Networks Topology Structure and Models

7.1.1 Basic Concepts of Network Topology Structure

(1) Graphical Representation of networks

A concrete network can be abstracted as a graph $G = (V, E)$ composed of point set V and edge set E. The number of nodes is $N = |V|$, and the number of edges is $M = |E|$. Each edge in E has a pair of points in V corresponding to it. If any pair of points (i, j) corresponds to the same edge (j, i), the network is called undirected network, otherwise it is called directed network. If each edge is given a corresponding weight, then the network is called weighted network, otherwise it is called unweighted network. Of course, an unauthorized network can also be regarded as an equal network with weights of 1 for each edge. In addition, a network may contain many different types of nodes. For example, in a social network, the familiarity of two people can be represented by weight, while different types of nodes can represent people of different nationalities, regions, ages, gender, and income. This book focuses on undirected weightless networks and assumes that there are no multiple edges and self-rings (that is, there is at most one edge between any two nodes, and there is no same node as the starting and ending edges). In graph theory, graphs without multiple edges and self-rings are called simple graphs.

(2) **Average path length**

The distance d_{ij} between two nodes i and j in the network is defined as the number of edges on the shortest path connecting the two nodes. The maximum distance between any two nodes in a network is called the diameter of the network, which is denoted as D, that is

$$D = \max_{i,j} d_{ij}.$$

© Tsinghua University Press 2020
X. Liang, *Social Computing with Artificial Intelligence*,
https://doi.org/10.1007/978-981-15-7760-4_7

The average path length L of the network is defined as the average of the distance between any two nodes, that is,

$$L = \frac{1}{\frac{1}{2}N(N+1)} \sum_{i \geq j} d_{ij}.$$

N is the number of network nodes. The average path length of the network is also called the characteristic path length of the network. In order to facilitate mathematical processing, the average path calculation includes the distance from the node to itself (that is, the distance is 0). The average path length of a network with N nodes and M edges can be determined by a breadth-first search algorithm with time magnitude $O(MN)$.

(3) Clustering coefficient

In the Friendship Network, two of a person's friends are likely to be friends with each other, which is called the clustering characteristics of the network. Generally, assuming that a node i in a network has a k_i edge and connects it to other nodes, the k_i node is called the neighbor of node i. Obviously, there may be at most $k_i(k_i - 1)/2$ edges between these k_i nodes. The ratio of the actual number of edges E_i and the total number of possible edges $k_i (k_i - 1)/2$ between k_i nodes is defined as the clustering coefficient C_i of node i. Namely

$$C_i = 2E_i/(k_i(k_i - 1)).$$

From the point of view of geometric characteristics, an equivalent definition of the upper formula is

$$C_i = \frac{\text{number of triangles connected with node } i}{\text{number of triads connected with node } i}.$$

wherein the triplet connected to the node i refers to three nodes including the node i, and at least two sides from the node i to the other two nodes exist.

The clustering coefficient C of the whole network is the average value of the clustering coefficient C_i of all nodes i. Obviously, $0 \leq C \leq 1$. $C = 0$ if and only if all nodes are isolated nodes, there is no connection edge; $C = 1$, if and only if the network is globally coupled, that is, any two nodes in the network are directly connected. In fact, in many types of networks (such as social networks), the probability that your friend's friend is also your friend tends to be a nonzero constant with the increase of network size. This means that these actual complex networks are not completely random, but to a certain extent have the characteristics similar to the social network of "things clustering, people clustering."

(4) Degree and degree distribution

Degree is a simple and important concept in the attributes of individual nodes. The degree k_i of node i is defined as the number of other nodes connected to that node.

The degree of a node in a directed network is divided into out-degree and in-degree. Node outgoing refers to the number of edges pointing from the node to other nodes, and node inbound refers to the number of edges pointing from other nodes to the node. Intuitively, the greater the degree of a node, the more "important" it is in a sense. The average value of the degree k_i of all nodes in the network becomes the average degree of the network, which is marked as <k>. The degree distribution of nodes in the network can be described by the distribution function $P(k)$. $P(k)$ denotes the probability that the degree of a randomly selected node is exactly K. Regular lattices have a simple degree sequence: because all nodes have the same degree, the degree distribution is delta distribution, which is a single peak. Any randomization tendency in the network will widen the shape of the spike. The degree distribution of complete random network is similar to Poisson distribution, and its shape decreases exponentially away from the peak value <K>. This means that when $k > k$, the node with degree K actually did not exist. Therefore, such networks are also called homogeneous networks.

A large number of studies have shown that the degree distribution of many real networks is obviously different from that of Poisson distribution. In particular, the degree distribution of many networks can be better described by power law form $P(k)$ $\propto k^{-\gamma}$. The power law distribution curve is much slower than the Poisson exponential distribution curve.

Power law distribution is also called scale-free distribution. The network with power law distribution is also called scale-free network. This is because the power law distribution function has the following scale-free properties.

Scale-free property of power law distribution function: consider a probability distribution function $f(s)$, if there is a constant b for any given constant a, the function $f(x)$ satisfies the following "scale-free condition,"

$$f(ax) = bf(x).$$

So, there must be (assume $f(1)f'(1) \neq 0$),

$$f(x) = f(1)x^{-\gamma}, \quad \gamma = -f(1)/f'(1).$$

In other words, the power law distribution function is the only probability distribution function that satisfies the scale-free condition. The derivation of properties is omitted in this book.

In a large scale-free network with proper power law distribution (usually $2 \leq \gamma \leq 3$), the degree of most nodes is relatively low, but there are a few nodes with relatively high degree. Therefore, this kind of network is also called inhomogeneous network, and those relatively high degree nodes are called "hub" of the network. For example, the U.S. highway network can be approximated as a uniform network, because it is impossible to tell hundreds of roads to pass through the same city; while the U.S. air network can be seen as a scale-free network, most airports are small airports, but there are a small number of very large airports connecting many small airports, such as Chicago, Dallas, Atlanta, and New York Airport.

Another way to represent degree data is to draw cumulative degree distribution function,

$$P_k = \sum_{k'=k}^{\infty} P(k').$$

It represents the probability distribution of nodes whose degree is not less than k.

If the degree distribution is a power law distribution, i.e., $P(k) \propto k^{-\gamma}$, then the cumulative degree distribution function conforms to the power law of power law exponent $\gamma - 1$,

$$P_k = \sum_{k'=k}^{\infty} k'^{-\gamma} \propto k^{-(\gamma-1)}.$$

If the degree distribution is exponential, i.e., $P(k) \propto e^{-k/\kappa}$, where $\kappa > 0$ is a constant, then the cumulative degree distribution function is exponential and has the same index,

$$P_k \propto \sum_{k'=k}^{\infty} \exp^{-k'/\kappa} \propto \exp^{-k/\kappa}.$$

Power law distribution corresponds to a straight line in logarithmic coordinate system and exponential distribution corresponds to a straight line in semilogarithmic coordinate system. Therefore, power law and exponential distribution can be easily identified by using logarithmic coordinate and semilogarithmic coordinate, respectively.

7.1.2 Regular Networks and Random Graphs

(1) Rule Network

This section introduces three most common rule networks: globally coupled network, nearest neighbor-coupled network, and star-coupled network.

First, in a globally coupled network, any two points are directly connected by edges. Therefore, in all networks with the same number of nodes, the globally coupled network has the smallest average path length $L_{gc} = 1$ and the largest clustering coefficient $C_{gc} = 1$. Although the global-coupled network model reflects the clustering and small world nature of the network in many centuries, its limitations as a practical network model are obvious. For example, a globally coupled network with N points has $N(N - 1)/2$ edges, but most large real networks are sparse, and the number of edges is usually at most $O(N)$ rather than $O(N^2)$.

Secondly, the nearest neighbor-coupled network is another sparse regular network which has been extensively studied, in which each node is only connected to its neighbors. The nearest neighbor coupled network with periodic boundary conditions consists of N nodes enclosed in a ring, where each node is connected to its left and right $K/2$ neighbor points, where K is an even number. For larger K values, the clustering coefficients of the nearest neighbor coupled network are

$$C_{nc} = \frac{3(K-2)}{4(K-1)} \approx \frac{3}{4}.$$

Therefore, such a network is highly clustered. However, the nearest neighbor coupled network is not a small world network. On the contrary, for a fixed K value, the average path length of the network is

$$L_{nc} = \frac{N}{2K} \to \infty (N \to \infty).$$

This can help explain from one side why it is difficult to implement dynamic processes (such as synchronization) that require global coordination in such a locally coupled network.

Thirdly, the star-coupled network has a central point, and the other N-1 points are only connected with this central point, but they are not connected with each other. The average path length of star network is

$$L_{star} = 2 - \frac{2(N-1)}{N(N-1)} \to 2(N \to \infty).$$

The clustering coefficients of star network are

$$C_{star} = \frac{N-1}{N} \to 1(N \to \infty).$$

Star network is a special kind of network. It is assumed that if a node has only one neighbor node, the clustering coefficient of the node is defined as 1. In some literatures, the clustering coefficient of the node with only one neighbor node is defined as 0. If defined in turn, the clustering coefficient of the star network is 0.

(2) Random Graph

Contrary to completely regular networks, totally random networks, one of the typical models is the ER random graph model [1] studied by Erdos and Renyi. Suppose that a large number of buttons $N > 1$ are scattered on the ground and each pair of buttons is tied to a line with the same probability P. In this way, we can get an example of ER random graph with N points and about $p_{N(N-1)/2}$ edges where (a) $P = 0$, given 10 outliers and (b) –(d) random graphs generated with connection probabilities $P = 0.1$, $P = 0.15$, and $P = 0.25$, respectively.

One of the main research topics of random graph theory is: what is the probability p when random graphs will produce some special attributes.

Erdos and Renyi systematically studied the relationship between the properties of ER immediate graph (such as connectivity) and probability p when $N \to \infty$. They adopted the following definitions:

If the probability of generating an ER random graph with property Q is 1 when $N \to \infty$, then almost every ER random graph has property Q.

The most important discovery of Erdos and Renyi is that ER random graphs have emergent or phase transition properties as follows.

Many important properties of ER random graphs emerge suddenly. That is to say, for any given probability p, almost every graph has some property Q (e.g., connectivity), or almost every graph does not have this property.

For example, for the above button network, if you pick up a button, how many buttons will be picked up? The results show that if the probability p is greater than a certain threshold $p_c \propto (\ln N)/N$, then almost every random graph is connected, that is, you randomly pick up a button will pick up almost all the buttons on the ground.

The average degree of ER random graph is $\langle k = P(N - 1) \approx pN \rangle$. Let L_{ER} be the average path length of ER random graph. Intuitively, for a randomly selected point in ER random graph, there are about $\langle k \rangle^{L_{ER}}$ other points in the network whose distance is equal to or very close to L_{ER}. Therefore, $N \propto \langle k \rangle^{L_{ER}}$, that is $L_{ER} \propto \ln N / \ln$ <k>. The characteristic that the average path length is a logarithmic growth function of network scale is a typical small world feature. Because the value of $\ln N$ increases slowly with N, even large networks can have very small average path length.

The probability of connection between two nodes in ER random graph is p, regardless of whether they have the same neighbor nodes or not. Therefore, the clustering coefficient of ER random graph is $C = p = \langle k \rangle / N < 1$, which means that large-scale sparse ER random graph has no clustering characteristics. In reality, complex networks generally have obvious clustering characteristics. That is to say, the clustering coefficients of real complex networks are much higher than those of ER random graphs of the same scale.

If the average degree <k> of a fixed ER random graph is unchanged, for sufficiently large N, the degree distribution of ER random graph can be expressed by Poisson distribution because the appearance of each edge is independent or not,

$$p(k) = \binom{N}{k} p^k (1 - p)^{N-k} \approx \frac{\langle k \rangle^k \exp^{-\langle k \rangle}}{k!}.$$

For fixed k, when N tends to infinity, the final approximate equation is exact. Therefore, ER random graph is also called "Poisson random graph."

Although ER random graph has obvious defects as a model of actual complex network, in the last 40 years of the 20th century, ER random graph theory has been the basic theory of complex network topology. Some of the basic ideas are still very important in the current complex network theory research. People can extend ER random graph from many angles to make it closer to the real network. One of the natural generalizations is the generalized random graph with any given distribution. Given a degree distribution $P(k)$, it represents the proportion of nodes with a degree of K in the network. Based on this distribution, a network composed of N nodes with $k_i (i = 1, 2, ..., N)$degree sequences can be generated according to the same probability. The set of these network models is called the configuration model.

7.1.3   Small World Network Model

As a transition from a completely regular network to a completely random graph, Watts and Strogtz [2] introduced an interesting small world network model in 1998, which is called WS small world model. The construction algorithm of WS small world model is as follows:

WS Small World Model Construction Algorithms:

1. Starting from the rule graph: Consider a nearest neighbor coupling network with N points, which is surrounded by a ring, in which each node is connected with its neighboring $K/2$ nodes, K is even.
2. Randomized reconnection: Reconnect each edge of the network randomly with probability p, that is, one endpoint of the edge remains unchanged, while the other endpoint is a randomly selected node in the network. It stipulates that there can be at most one edge between any two different nodes, and each node cannot be connected with itself.

In the above model, $P = 0$ corresponds to a completely regular network and $P = 1$ corresponds to a completely random network. By adjusting the value of p, the transition from a completely regular network to a completely random network can be controlled.

The clustering coefficient $C(p)$ and the average path length $L(p)$ of the network model obtained by the above algorithm can be regarded as functions of the reconnection probability P. The following figure shows the relationship between the clustering coefficient and the average path length of the network with the probability of reconnection P (two values are normalized along the way). A fully regular nearest neighbor coupled network (corresponding to $P = 0$) is highly clustered ($C(0) \approx 3/4$), but the average path length is large, ($L(0) \approx N/2K > 1$). When p is small ($0 < p < 1$), the local attributes of the reconnected network are not different from those of the original planning network, so the clustering coefficient of the network does not change much, ($C(p) \propto C(0)$), but the average path length decreases rapidly ($L(p) < L(0)$). Such networks with short average path length and high clustering coefficient become small world networks.

The randomization process in WS small world model construction algorithm may destroy the connectivity of the network. Another small world model that has been studied more is the NW small world model, which was subsequently proposed by Newman and Watts. This model is obtained by substituting "randomized reconnection" in WS small world model construction with "randomized edge addition." The specific construction algorithm is as follows,

NW Small World Model Construction Algorithms:

1. Starting from the rule graph: Consider a nearest neighbor coupling network with N points, which is surrounded by a ring, in which each node is connected with its neighboring $K/2$ nodes, K is even.
2. Randomized edge addition: A probability p is used to add an edge between randomly selected pairs of nodes. Among them, any two different nodes can only have one side at most, and each node cannot be connected with its own side.

In the NW small world model, $P = 0$ corresponds to the original nearest neighbor coupling network, and $P = 1$ corresponds to the global coupling network. In theoretical analysis, NW small world model is simpler than WS small world model. When P is small enough and N is large enough, NW small world model is essentially equivalent to WS small world model.

The small world network model reflects a characteristic of the network of friends, that is, most people and friends are neighbors living on the same street or colleagues working in the same unit. On the other hand, some people are friends who live far away, even far away from other countries. This situation corresponds to the remote connection generated by reconnection in WS small world model or by joining wires in NW small world model [3–5].

Here are some statistical properties of the small world network model.

(1) Clustering Coefficient

The clustering coefficients of WS small world networks are as follows,

$$c(p) = \frac{3(K - 2)}{4(K - 1)}(1 - p)^3.$$

The clustering coefficients of NW small world networks are

$$c(p) = \frac{3(K - 2)}{4(K - 1) + 4Kp(p + 2)}.$$

(2) Average Path Length

So far, there is no exact analytical expression for the average path length L of WS small world model. However, the following formula can be obtained by using renormalization group method,

$$L(p) = \frac{2N}{K} f(NKp/2),$$

where $f(u)$ is a universal scaling function, satisfying

$$f(u) = \begin{cases} \text{constant}, u < 1 \\ (\ln)u/u, u > 1 \end{cases}.$$

Newman et al. gave the following approximate expressions based on the mean-field method,

$$f(x) \approx \frac{1}{2\sqrt{x^2 + 2x}} \arctan h\sqrt{\frac{x}{x+2}}.$$

But up to now, there is no exact explicit expression of $F(u)$.

(3) Degree Distribution

In the NW small world model based on the "randomized plus edge" mechanism, each node has a degree of at least K. Therefore, when $k > K$, the probability that a randomly selected node has a degree of k is

$$p(k) = \binom{N}{k-K}\left(\frac{Kp}{N}\right)^{k-K}\left(1 - \frac{Kp}{N}\right)^{N-k+K}.$$

And when $k < K$, $P(k) = 0$.

For the WS small world model based on the "randomized reconnection" mechanism, when $k > K/2$,

$$p(k) = \sum_{n=0}^{\min(k-K/2,k/2)} \binom{K/2}{n}(1-p)^n p^{(K/2)-n} \frac{(pK/2)^{k-(K/2)-n}}{(k-(K/2)-n)!} \exp^{-pK/2}.$$

And when c, $P(k) = 0$.

Similar to ER random graph model, WS small world model is also a homogeneous network with approximately equal degrees of all nodes.

7.1.4   Scale-Free Network Model

A common feature of ER random graph and WS small world model is that the connectivity distribution of networks can be approximately expressed by Poisson distribution, which has a peak value at the degree average $<k>$ and then decays exponentially rapidly. This means that when $K > \langle k \rangle$, the node with degree K hardly exists. Therefore, such networks are also called exponential networks. Another important discovery in the study of complex networks is that many complex networks, including Internet, WWW, and metabolic networks, have power law connectivity distribution functions. Because the connectivity of nodes in such networks has no obvious characteristic length, it is called scale-free network.

In order to explain the generation mechanism of power law distribution, Barabasi and Albert proposed a scale-free network model [14], which is now called BA model. They argue that many previous network models did not take into account the following two important characteristics of the actual network:

1. Growth characteristics: The scale of the network is constantly expanding. For example, a large number of new research articles are published every month, while a large number of new Web pages are generated every day on WWW.

2. Preferential attachment: New nodes are more likely to connect with "big" nodes with high connectivity. This phenomenon is also called "rich get richer" or "Matthew effect." For example, newly published articles tend to cite some important documents that have been widely cited, and hypertext links on new personal homepages are more likely to point to famous Web sites such as Yahoo and Google.

Based on the growth and preferential connection characteristics of the network, the construction algorithm of BA scale-free network model is as follows:
Construction Algorithm of BA Scale-free Model:

1. Growth: Starting from a network with m_0 node, each time a new node is introduced and connected to m existing nodes, here $m \leq m_0$.
2. Priority connection: the probability of a new node connecting to an existing node i, $\prod i$, degree k_i of node I, and degree k_j of node j satisfy the following relationship,

$$\prod i = \frac{k_i}{\sum_j k_j}.$$

After t-step, the algorithm generates a network with $N = t + m_0$ nodes and m_t edges. The following figure shows the evolution of BA network when $m = m_0 = 2$. There are two nodes in the initial network. Each additional node is connected to two existing nodes in the network according to the priority connection mechanism.

(1) Clustering coefficient

The clustering coefficients of BA scale-free networks are as follows,

$$c = \frac{m^2(m+1)^2}{4(m-1)}\left[\ln\left(\frac{m+1}{m}\right) - \frac{1}{m+1}\right]\frac{[\ln(t)]^2}{t}.$$

This shows that similar to ER random graph, BA scale-free network has no obvious clustering characteristics when the network scale is large enough.

(2) Degree distribution

At present, there are three main methods to study the degree distribution of BA scale-free networks: continuum theory, master equation method and rate equation method. The results of the three methods are the same. The principal equation method and the rate equation method are equivalent. The results obtained by the main equation method are described below.

$P(k, t_i, t)$ is defined as the probability that the degree of node i added at time t_i is exactly K. In BA model, when a new node is added to the system, the probability of reading increase 1 of node i is $m \prod i = k/2t$, otherwise the degree of the node remains unchanged. Thus, the following recursive formula is obtained,

$$P(k, t_i, t+1) = \frac{k-1}{2t} P(k-1, t_i, t) + \left(1 - \frac{k}{2t}\right) P(k, t_i, t).$$

The degree distribution of the network is

$$p(k) = \lim_{t \to \infty} \left(\frac{1}{t} \sum_{t_i} P(k, t_i, t)\right).$$

It satisfies the following recursive equation,

$$P(k) = \begin{cases} \frac{k-1}{k+2} P(k-1), & k \geq m+1, \\ \frac{2}{m+2}, & k = m. \end{cases}$$

The degree distribution function of BA network is obtained as follows,

$$P(k) = \frac{2m(m+1)}{k(k+1)(k+2)} \propto 2m^2 k^{-3}.$$

This shows that the degree distribution function of BA network can be approximated by the power law function with power exponent 3. It should be pointed out that there are still some different views on the construction of BA scale-free network model and the rigor of its theoretical analysis.

7.1.5 Local World Evolutionary Network Model

In the study of the World Trade Web, it is found that the global priority connection mechanism does not apply to those countries which only have trade relations with a few (less than 20) countries. In this trade network, each node represents a country; if there is trade relationship between two countries, there is a connection between the corresponding two nodes. Research shows that many countries are committed to strengthening economic cooperation and trade relations with countries within their respective regional economic cooperation organizations. These organizations include the European Union (EU), ASEAN, and North American Free Trade Area (NAFTA). In the world trade network, priority connection mechanism exists in some regional economies. Similarly, in the Internet, computer networks are organized and managed based on the domain-router structure. A host is usually only connected to other hosts in the same domain, while a router is connected to other routers on behalf of its internal host. The priority connection mechanism is not for the whole network, but is effective in each node's local-world. Even in people's organizations, everyone actually lives in their own local world. All of these show the existence of the local world in many practical complex networks.

In many real networks, because of the existence of local-world connectivity, each node has its own local-world, so only the local-connection information of the whole network is occupied and used. The local-world evolving network model [15] is used to describe this situation. The construction algorithm of the model is as follows:

Local World Evolution Model Construction Algorithms:

1. Growth: There are m_0 nodes and e_0-edges at the beginning of the network. Each time a new node and the attached m-strip edge are added.
2. Local World Priority Connection: Randomly select M nodes ($M \geq m$) from the existing nodes of the network as the local world of the newly joined nodes. New nodes are added according to the priority connection probability,

$$\prod \text{local}(k_i) = \prod{}' (i\text{LW}) \frac{k_i}{\sum_j \text{local}(k_j)} \equiv \frac{M}{m_0 + t} \frac{k_i}{\sum_j \text{local}(k_j)},$$

to select m nodes connected to the local world, where LW consists of newly selected M nodes.

At each moment, the newly joined nodes are selected from the local world according to the priority connection principle to connect m nodes, rather than from the whole network like BA scale-free model. The rules for constructing a local world of a node depend on the actual local connectivity. The above model only considers the simple case of random selection.

Obviously, at time $t, m \leq M \leq m_0 + t$. Therefore, there are two special cases in the above-mentioned local-world evolutionary network model: $M = m$ and $M = t + m_0$.

(1) Special case A: $M = m$

At this time, the newly added nodes are connected to all the nodes in their local world, which means that the priority connection principle has virtually no longer worked in the process of network growth. This is equivalent to a special case in BA scale-free network model where only growth mechanism is preserved, and no preferential connection is given. At this time, the change rate of the degree of the ith node is.

$$\frac{\partial k_i}{\partial t} = \frac{m}{m_0 + t}.$$

The distribution of network degree obeys exponential distribution,

$$P(k) \propto \exp^{-\frac{k}{m}}.$$

(2) Special case B: $M = t + m_0$

In this special case, the local world of each node is actually the whole network. There-fore, the local world model is completely equivalent to the BA scale-free network model at this time.

7.2 Online Social Networks Community Discovery Models

7.2.1 Basic Concepts of Community

A precise definition of what an "online social networks community" really is does not exist yet. One of the most widely accepted and used definition of "community" is that Newman and Girvan [16]: A community is a sub-graph containing nodes which are more densely linked to each other than to the rest of the graph or equivalently, a graph has a community structure if the number of links into any sub-graph is higher than the number of links between those sub-graphs.

The so-called community detection refers to the mapping $G = G(V, E)$, in which $nc(\geq 1)$ communities are identified,

$$C = \{C_1, C_2, ..., C_{nc}\}.$$

Let the vertex set of each community form a cover of V.

Community is the common characteristic of all kinds of online social networks. According to the definition of Girvan and Newman, members will form closely related groups in social networks. Generally speaking, the interaction between indi-viduals within a group is more frequent than that between individuals outside the group [6, 7]. If the vertex set of any two communities is empty, then C is called disjoint communities; otherwise, it is called overlapping communities.

7.2.2 Online Networks Community Discovery-Related Technologies Model

In order to study the community structure characteristics of online social networks, researchers have proposed many algorithms and models for community discovery [8–10]. This section divides these technologies into static community discovery and dynamic community discovery, respectively. Among them, static community discovery means that the nodes and edges in the community network are in a static state and will not change with time. Dynamic community discovery is a method of extending the static community discovery model to the time axis to analyze the community networks that update nodes and edges over time.

7.2.2.1 Static Community Discovery Model

In static community discovery, there are mainly spectral analysis, based on the idea of connecting edge-median, information centrality, and objective function. Researchers have proposed some representative models, such as segmentation model, hierarchical clustering model in sociology, W-H model, and GN model.

(1) Graph segmentation model

Generally, graph segmentation community discovery model is regarded as a community discovery technology in the field of computer science. The typical application of graph segmentation is parallel computing in the field of computer. Scholars have designed many heuristic algorithms and found some better solutions. Most graph segmentation methods are based on iterative partitioning: firstly, the whole graph is decomposed into two optimal sub-graphs, and then, the two sub-graphs are decomposed into two optimal sub-graphs, and the same processing is repeated until a sufficient number of sub-graphs are obtained, the most famous of which is the Kernighan–Lin model [11].

Kernighan–Lin model: The network is decomposed into two communities of known size through a heuristic process based on greedy optimization. In this model, a gain function Q is introduced to partition the network, which is defined as the difference between the number of edges falling inside the community and the number of edges falling between the two communities, and then, a partitioning method to maximize Q is found.

(2) W–H model

The biggest disadvantage of the traditional graph segmentation method is that it can only divide the network equally at a time. If a network is to be divided into more than two communities, the method must be repeated many times for the divided sub-communities. To solve this problem, Wu and Huberman [12] proposed a fast segmentation method based on resistance network voltage spectrum, called W-H model. The basic idea of W-H model is that if two nodes which are not in the same community are regarded as source node (voltage 1) and terminal node (voltage 0), and each side is regarded as a resistance with resistance value of 1, then the voltage value between nodes in the same community should be close. Therefore, as long as the source node and the end node are found by the correct method and a suitable voltage threshold is selected, the correct community structure can be obtained.

An important feature of W-H model is that it can find the whole community of a known node without considering the whole network community structure, without calculating all the communities. This feature can be well applied in WWW, Web search engine, and other large-scale networks. The disadvantage of W-H model is that it is difficult to apply this algorithm if some information about the structure of network community is not known beforehand.

(3) Split hierarchical clustering

Because the number of communities in a network is not known beforehand, and the number of nodes in each community is unknown, community discovery is a more difficult problem than graph segmentation. In addition, the community structure of the network usually presents hierarchical characteristics. A community can be further divided into several sub-communities, which also increases the difficulty of community discovery. In order to solve these problems, sociologists put forward a hierarchical clustering method [13, 14].

The idea of hierarchical clustering method in community network analysis is closer to the idea of community structure. The purpose is to find out the community structure in social network according to various criteria to measure the similarity degree and the close degree of connection between nodes. Firstly, the splitting clustering algorithm divides the nodes into several disjoint sets and then further divides each set into smaller sets until each set contains few or even only one node. The key is how to divide a network into several parts. Some partitioning methods, such as block model, spectral clustering model, and implicit spatial model, can be repeatedly used to discover communities as smaller sets.

The most famous algorithm of split hierarchical clustering is GN model [15, 16]. In this model, Girvan and Newman introduced the concept of edge median. The edge median of one edge is defined as the number of paths passing through the edge in the shortest path between all pairs of nodes. In a graph with m edges and n nodes, the time required to calculate the median of each edge is $O(mn)$. The basic step of GN algorithm is to calculate betweenness on the edge of the network, delete the edge with the highest betweenness value, and recalculate betweenness on all edges until all edges are deleted.

Generally, the GN model can also be regarded as a splitting method. Unlike the general splitting method, the GN model does not look for the weakest pairs of nodes and then delete the edges between them, but for the most "between" edges and delete it.

Aggregated clustering algorithm starts from the basic community and aggregates it into larger communities according to some criteria. One of the criteria is modularity [17, 18]. If the merger of the two communities can achieve higher overall modularity, then merge them. Start with each node as a community, and then merge until the modularity cannot be improved.

Clauset, Newman, and Moore proposed a new aggregation model for computing and updating network modularity using data structures such as heaps, which is called CNM model [19, 20]. By calculating the whole network, the CNM model combines two points i and j into two points with the greatest possible increment of modularity ΔQ_{ij} in a community and merges them into one community. Then, it regards these two points as one point and repeats the steps above until all points are grouped into one community and then chooses the largest division of Q as the result of partition. The time complexity of CNM algorithm is $O(m\log n)$.

Louvain method [21] is another hierarchical clustering model. This model is a greedy algorithm based on Q. The time complexity of the algorithm is low. When the network is sparse, it can almost achieve linear time complexity, and the community discovery results are excellent. It has been introduced by social computing tools such as network X and i graph. The main calculation steps of the model are as follows:

Step 1: Assume that each node in the network is a separate community in the initial case.

Step 2: For each node, the modularity change Q is calculated when the node joins the neighboring node community, and the node is merged into the community with the largest modularity increment. If the change of Q is less than 0, then the node is reserved in the current community; repeat step 2 until all points are traversed, and Q achieves the optimum.

Step 3: Consider the current node in a community as a new node, and repeat step 2 for these new nodes until Q is reached.

Step 4: Repeat steps 4, 3, and 2 until Q_{max} is reached.

4. **Clustering model based on partition**

K-means model is a typical partition-based clustering model. The flow of the model is described as follows: first, K objects are randomly selected, each object is regarded as the initial center of the community; for each remaining object, it is assigned to the cluster represented by the nearest center according to some distance measure with each community center. Then, the new centers of each community are recalculated, and the process is repeated until the criterion function converges. This clustering model is scalable and efficient. There are many improved or optimized clustering algorithms based on this idea. However, due to the need to set the number of communities in advance, there are some limitations.

Although researchers have given a variety of methods for community discovery, there is no commonly used method at this stage. This book summarizes the above static community discovery models as shown in the following table, so as to facilitate researchers to choose models for different social computing problems (see Table 7.1).

7.2.2.2   Dynamic Community Discovery Model

Online social networks are constantly changing. With the change of users' interests, the development of news and events in society, and the change of influence, if we only study the community discovery of the network at a certain time point, we will lose a lot of information of the network. Chakrabarti found that the community structure of dynamic networks is smooth in a short time, which is called "short smoothness" of dynamic communities. On the one hand, "short smoothness" means that the network structure of the next moment in the network is very similar to the network structure of the current moment. The network structure of the future moment can be found based on the network structure of the current moment. On the other hand, "short smoothing" is a basic feature of dynamic networks and can be used as one of the evaluation indicators to evaluate the results of dynamic community discovery.

At present, the main research contents of dynamic social network community discovery include three aspects:

Table 7.1 Static community discovery model comparison summary

Model name	Advantage	Disadvantage	Time complexity
W-H model	Able to find an entire community without considering the entire network community structure	Difficult to apply when some information about the structure of the network community is not known beforehand	$O(n)$
GN model	Do not need to specify the number and size of the communities to be decomposed beforehand, and the accuracy is high	Slow	$O(m^2 n)$
CNM model	Approach linear complexity, using heap structure to store data saves storage space	In order to find modularity as the goal, ignoring the practical significance of the network	$O(n \log^2 n)$
K-means model	Fast, scalable, practical, and suitable for the core stable community structure	Need a priori knowledge (number of associations), susceptible to noise	$O(nkt)$

Firstly, according to the community structure of discrete time point networks, the evolution of communities and communities in networks is analyzed, such as the generation, change, integration, division, and disappearance of communities.

Secondly, according to the community discovery of the network at different times, mining the stable structure and core layer of the network communities.

Thirdly, according to the historical information (usually referring to the time step $t-1$ in network graph and the corresponding community discovery) and the topological structure of the time step t network, the community structure of the time step t network is inferred.

This section mainly introduces the existing dynamic community discovery model.

(1)   Extending static method to dynamic method

Mucha [22] extended the calculation method of modularity Q in the static community evaluation standard to the time axis by using Laplacian dynamic model and generates a new dynamic community evaluation standard. Based on this evaluation criterion, an empty model is proposed, which can deal with different time span, different scale, and different types of communities, especially those with stable core structure.

Chakrabarti et al. [23] will use modularity Q to evaluate community discovery results at point t as part of the goal. Taking the change of community structure from t to $t+1$ as another part of the objective function, the objective function is formed. In the process of evolution, the change cost of evolution is the smallest, and the modularity is the largest,

$$sq(C_t, M_t) - cp * hc(C_{t-1}, C_t).$$

Chi uses similar methods to develop graph segmentation into dynamic community discovery and implements spectral clustering with smoothness constraints. Kim et al. [24] used cost embedding technology to measure the cost of community change in nano-communities and established a density-based dynamic community discovery method.

(2) **Hidden space-based approach**

Hidden space-based approach regards community as a kind of hidden space in the network. In the structure of hidden space, the nodes with closer distance are easier to establish connections than those with long distance. Compared with the extended static method, the main difference of this method is that the extended static method mostly aims at finding the results with less spatiotemporal variation, while the implicit space-based method aims at maximizing the posteriori probability [25, 26].

Sarkar et al. [27] used the exponential distribution probability to analyze the network results on each time slice, used the Gauss distribution to establish the transfer model, used the multidimensional scaling to initialize the spatial location, used the kernel function to calculate the approximation, established rules to find the individual location, and then monitored the network changes.

Xu et al. used Dirichlet process to establish a prior probability distribution of community network, combined with hidden Markov model, established a Bayesian learning model, introduced short smoothing hypothesis, established a community attenuation factor, and used transformation matrix to transform the relationship between communities at different time points.

(3) **Block-based approach**

Block model can show the block structure between nodes with similar attributes, which is suitable for community discovery. In this method, the relational attributes of nodes are represented as matrices, and the communities are found by rearranging the matrices. There are three main characteristics of block model, structural equivalence: two nodes have the same structure in the network; rule equivalence: two nodes have the same location under some rules; random equivalence: two nodes can exchange positions with a certain probability, which has little impact on the network.

The Chinese restaurant process (CRP) is a typical block model. It is often used as a node clustering model: a user will choose to sit at a table or a new table according to the situation of someone at the table.

(4) **Incremental community discovery methods**

Online social networks have a large amount of data and strong dynamic. It is an urgent need for practical application to quickly and accurately discover the hidden community structure and obtain the evolution information of the hidden structure. To solve the problem of algorithm efficiency, many incremental dynamic community discovery models have been proposed. On the basis of identifying the community structure at the previous moment, only the part of the changed network structure is recalculated, while the rest of the structure remains unchanged, so as to improve the computational efficiency.

Current incremental strategies are mainly divided into physical-based incremental strategies and graph-based incremental strategies.

1. Incremental strategy based on physical principles regards the network as a complex physical world. Inspired by Newton's law of universal gravitation, Yang et al. divides the relationship between network nodes into two kinds: attraction and exclusion. Through iterative incremental calculation, the attraction of the formed community becomes more and more strong, and the exclusion of the edges of the communities becomes more and more strong. Finally, the connecting edges break up and form a partitioned community structure.

2. Incremental dynamic community discovery method based on graph features usually uses static community discovery algorithm to discover the community structure on the initial time snapshot, then identifies the structural changes in the subsequent network snapshot, and calculates the structure of these changes, so as to avoid recalculating the whole network. The main incremental strategies are based on graph segmentation feature, spectral feature, and matrix decomposition.

Under the condition that the community structure at $T - 1$ time is known, through setting and identifying new vertices and changing community attribution conditions, Shan et al. only analyzed the incremental correlation nodes to get the community structure at t time, avoiding rediscovering the community and improving the computational performance. However, this method does not consider the increase and disappearance of nodes in the process of network change, and the number of communities in the calculation process is kept unchanged by default. The basic idea of this algorithm is to define the degree of dependence of nodes on community, and aff represents the degree of dependence of vertices on community, that is, the ratio of edges between vertices V and vertices in community I to all edges connected with V,

$$\text{aff}_{v,i} = \frac{\{u | u C_{t-1}, (u, v) E_t\}}{d_v}.$$

In the following period of time, a vertex is defined to change community attribution conditions as follows,

$$\frac{\text{aff}_{v,j} - \text{aff}_{v,i}}{\text{aff}_{v,i}} > \varepsilon.$$

Hang proposed [28] incremental spectral clustering method, which takes into account the increase, deletion of nodes, and the change of similarity between nodes. By evaluating the impact of changes in nodes and edges on network eigenvalues and eigenvectors, a transformed Laplace matrix is introduced to replace the matrix. The maximum eigenvector of the current time snapshot is calculated approximately by using the result of the eigenvector of the previous time point with the method of approximate eigenvector calculation, which reduces the reclustering. Compared with the static spectral clustering algorithm, the computing performance is improved; Thang et al. [29] divides the nodes into leader and follower roles according to the

Table 7.2 Dynamic community discovery model comparison summary

Method type	Advantage	Disadvantage	Time complexity
From static method to dynamic method	The model is mature and takes into account the "short smoothness" characteristics and modularity of the community	Redividing the community while considering the stability of the community structure	High
Bayesian model method	The results are better in modularity, and overlapping and non-overlapping communities can be found	Depending on prior knowledge, the number of communities needs to be continuously input	High
Block model method	Multiple population structures can be found	It is more suitable in the same probability distribution network. The actual network nodes have large differences, wide degree distribution and unsatisfactory model results	High
Incremental community discovery method	Using incremental information for community discovery keeps the consistency of community structure	Because only incremental information is considered, the accuracy of the model is not high enough compared with the static expansion method	Low

node degree characteristics for scale-free networks, designs incremental computing rules, finds dynamic communities, and gives the approximate ratio guarantee of the algorithm.

The advantages and disadvantages of the dynamic community discovery model and its algorithm examples in this section are summarized in the table below for readers' overview (see Table 7.2).

7.2.3   Evaluation Indicators of Community Discovery Model

(1) Modularity

In order to measure the quality of community discovery methods in real networks, Newman proposes a modularity method, which takes into account the distribution of node degrees. Modularity is used to measure the degree of density of links within a module by calculating the difference between the number of links within a community and the number of links within a community under random circumstances [30].

Given a network consisting of n nodes and m edges, between any two nodes v_i and v_j, the expected value of the edge is $k_i k_j / 2m$, where k_i and k_j are the v_i and v_j degrees of two nodes, respectively. Considering that an edge starts from the node v_i and randomly connects any node in the network, then it connects the node v_j with the probability of $k_j / 2m$. Because the degree of node v_i is k_i, that is to say, the node has k_i edges, so the expected value of the edge is $k_i k_j / 2m$ between two nodes v_i and v_j.

By calculating $A_{ij} - k_i k_j / 2m$, we can know the difference between the actual interaction between v_i and v_j of any node and the expected number of connections. Given the set C of a node, the community effect intensity can be defined as,

$$\sum_{i,jC} A_{ij} - k_i k_j / 2m.$$

If a network is divided into k communities, then the overall community effect intensity is defined as,

$$\sum_{l=1}^{k} \sum_{i,jC} A_{ij} - k_i k_j / 2m.$$

Modularity is defined as,

$$Q = \frac{1}{2m} \sum_{l=1}^{k} \sum_{i,jC} A_{ij} - k_i k_j / 2m.$$

Among them, the parameter $\frac{1}{2m}$ is introduced to normalize the Q value to $[-1, 1]$ interval.

The higher the value, the better the cohesion of the community is. By calculating the value, we can judge whether the community discovery algorithm obtained in the community discovery algorithm is good or bad. After the concept of modularity was put forward, it has been widely recognized. In addition, modularity has been applied to partition effect evaluation of algorithm and community structure detection of network combined with data mining algorithm.

Modularity is often expressed in another form,

$$Q = \frac{1}{2m} \sum_{ij} \frac{A_{ij} - k_i k_j}{2m} * \delta(C_i, C_j),$$

where

$$\delta(C_i, C_j) = \begin{cases} 1, C_i = C_j, \\ 0, \text{else.} \end{cases}$$

Based on the modularity of Newman and Girvan, many researchers have proposed modularity in different networks. For example, in the weighted network with W_{ij} edge, modularity can be expressed as,

$$Q = \frac{1}{2m} \sum_{ij} \left(W_{ij} - \frac{k_i k_j}{2m} \right) * \delta(C_i, C_j).$$

(2) **Interest cohesion index analysis (C^i)**

Interest cohesion means that the interests of users in the community should have the characteristics of aggregation. In order to better describe the characteristics of interest cohesion in the community, this paper introduces the interest cohesion index C^i, which measures the degree of user interest cohesion in the community, to measure the ratio between the similarity S of user interest in the community and the similarity of user interest in the whole network. Its definition is as follows,

$$C^i = \frac{\Sigma_{\text{number of } a} \Sigma S(W_{ij}), \text{all points pairs } W_{ij} \text{ in the same community } a}{\Sigma S(W_{ij}), \text{all points pairs } W_{ij} \text{ in the network}}.$$

Obviously, the higher the C^i value, the better the interest cohesion of the community as a whole, and the better the performance of the user community discovery algorithm in aggregating similar interest users.

(3) **AC cohesion index analysis (E^i)**

Similarly, communicative cohesion analysis refers to directional communication between users mostly concentrated within the community. In order to better describe the community cohesion of communication, this section refers to the definition of interest cohesion and uses the exchange cohesion index E^i to measure the ratio of the number of user exchanges within the community to the number of exchanges with all users on the network. Its definition is as follows,

$$E^i = \frac{\Sigma_{\text{number of } a} \Sigma \text{ all } W_{ij}\text{'s numbers of communications } C_{ij} \text{ in the same community } a}{\Sigma \text{ all } W_{ij}\text{'s numbers of communications } C_{ij} \text{ in the network}}$$

Obviously, the higher the E^i value, the better the cohesion of the communities found, and the better the performance of the user community discovery algorithm in aggregating frequent users.

References

1. Erdos P, Renyi A (2011) On the evolution of random graphs. Trans Am Math Soc 286(1):257–274
2. Watts DJ, Strogatz SH (1998) Collective dynamics of 'small-world' networks. Nature 440–442
3. Barabasi AL, Albert R (1999) Emergence of scaling in random networks. Science 286(5439):509–512
4. Li X, Chen G (2003) A local-world evolving network model. Phys A 328(1):274–286
5. Newman ME, Girvan M (2004) Finding and evaluating community structure in networks. Phys Rev E: Stat, Nonlin, Soft Matter Phys 69(2):026113
6. Traud A, Kelsic E, Mucha P (2011) Community structure in online collegiate social networks. SIAM Rev 53(3):526–543
7. Haythornthwaite C (2009) Social networks and online community. Social networks and online community. Oxford Handbooks
8. Cai H, Zheng V, Chang K (2018) A comprehensive survey of graph embedding: problems, techniques, and applications. IEEE Trans Knowl Data Eng 30(9):1616–1637
9. Tang L, Liu H (2010) Community detection and mining in social media. Morgan and Claypool Publishers, p 137
10. Clauset A, Newman ME, Moore C (2004) Finding community structure in very large networks. Phys Rev E: Stat, Nonlin, Soft Matter Phys 70(2):066111
11. Lim K H, Datta A. Finding twitter communities with common interests using following links of celebrities. International Workshop on Modeling Social Media, ACM Conference, pp 25–32
12. Huberman BA, Romero DM, Wu F (2008) Social networks that matter: twitter under the microscope, vol 14(1), p 2009. Social Science Electronic Publishing
13. Toujani R, Akaichi J (2018) Hybrid hierarchical clustering approach for community detection in social network. International Conference on Ubiquitous Intelligence and Computing
14. Li L, Gui X, An J (2015) Overlapping community detection algorithm based on fuzzy hierarchical clustering in social network. J Xian Jiaotong Univ 49(page:6):6–13
15. Girvan M, Newman ME (2002) Community structure in social and biological networks. Proc Natl Acad Sci 99(12):7821–7826
16. Sathiyakumari K, Vijaya MS (2016) Community detection based on Girvan Newman algorithm and link analysis of social media. Convention of the Computer Society of India. Springer, Singapore, pp 223–234
17. Ghaderi S, Abdollahpouri A, Moradi P (2017) On the modularity improvement for community detection in overlapping social networks. IEEE International Symposium on Telecommunications. IEEE, pp 540–546 (2017)
18. Bu Z, Zhang C, Xia Z (2013) A fast parallel modularity optimization algorithm for community detection in online social network. Knowl-Based Syst 50(3):246–259
19. Clauset A (2005) Finding local community structure in networks. Phys Rev E: Stat Nonlin Soft Matter Phys 72(2):026132
20. Seth S, Bhattacharyya D, Kim T (2016) CBACCN: constraint based community discovery in complex network, vol 9(23). Social Science Electronic Publishing, pp 973-4562
21. Forster R (2016) Louvain community detection with parallel heuristics on GPUs. In: IEEE International Conference on Intelligent Engineering Systems, pp 227–232
22. Mucha PJ, Richardson T, Macon K (2010) Community structure in time-dependent, multiscale, and multiplex networks. Science 328(5980):876
23. Chakrabarti D, Kumar R, Tomkins A (2015) Evolutionary clustering
24. Kim MS, Han J (2009) A particle-and-density based evolutionary clustering method for dynamic networks. Proc VLDB Endowment 2(1):622–633
25. Dong Z (2016) Dynamic community detection algorithm based on hidden Markov model. In: International Symposium on Advances in Electrical, Electronics and Computer Engineering
26. Liu D, Liu Z, Zhang X (2012) Community structure detection via hidden Markov random field model. Int J Adv Comput Technol

27. Zhang J, Philip SY (2019) Broad learning through fusions: an application on social networks. Springer, Berlin
28. Ning H, Xu W, Chi Y (2010) Incremental spectral clustering by efficiently updating the Eigensystem. Pattern Recogn 43(1):113–127
29. Dinh TN, Nguyen NP, Thai MT (2013) An adaptive approximation algorithm for community detection in dynamic scale-free networks. IEEE INFOCOM, pp 55–59
30. Newman ME (2016) Community detection in networks: modularity optimization and maximum likelihood are equivalent

Chapter 8
Social Network Propagation Mechanism and Online User Behavior Analysis

8.1 Network Propagation Mechanism and Dynamic Analysis

It is of great research value to study how information is disseminated in online social networks. For example, in the management of network public opinion, information dynamic transmission model can help people better understand the mechanism of message transmission and control the spread of false information; in the field of public health, controlling the spread of infectious diseases, with the help of transmission threshold theory, can help people better understand the mechanism of virus transmission and thus help to reduce infection. Therefore, this section will introduce the propagation threshold theory and the propagation dynamics model in complex networks.

8.1.1 Propagation Critical Value Theory for Complex Networks

In a typical communication model, individuals in a population are abstracted into several categories, each of which is in a typical state. Its basic states include: S (susceptible)—healthy state; I (infected)—infected state; and R (removed, refractory or recovered)—removed state (also known as immune state or restored state). The transition between these states is usually used to name different infectious models [1–3]. For example, If a susceptible population is infected, then comes back to health and has immunity, it is called the SIR model. And if the susceptible group is infected and returns to the susceptible state, it is called SIS model. Based on these node dynamics models, the critical value properties of virus propagation in complex networks have been studied. In this paper, we will mainly introduce the research work on the propagation critical theory in uniform networks (such as WS small-world networks) and

© Tsinghua University Press 2020
X. Liang, *Social Computing with Artificial Intelligence*,
https://doi.org/10.1007/978-981-15-7760-4_8

non-uniform networks (such as BA scale-free networks), associated and unrelated networks.

(1) Propagation Critical Value of Uniform Network

Suppose that the nodes in the network follow the SIS model of easy-to-dye (S) → infection (I) → easy-to-dye (S). Make the probability from easy-to-infect state γ and from infect state to easy-to-infect state δ. Define effective transmission rate λ as follows,

$$\lambda = \frac{\gamma}{\delta}.$$

Without losing generality, suppose $\delta = 1$, because this only affects the definition of the time scale of disease transmission.

When time t is defined, the density of infected nodes is $P(t)$. When time t tends to be infinite, the steady-state density of infected individuals is denoted as ρ. We can use the mean-field theory to analytically study SIS model. For this reason, three assumptions are given for uniform networks.

Homogeneity hypothesis: the degree distribution of homogeneous networks (e.g., ER random graph and WS small-world network) has a peak at the network average $\langle k \rangle$ and decreases exponentially when $k \prec \langle k \rangle$ and $k \succ \langle k \rangle$. Therefore, it is assumed that the degree k_i of each node in the network is approximately equal to $\langle k \rangle$.

Homogeneous mixing hypothesis: the intensity of infection is proportional to the density $\rho(t)$ of infected individuals. One of its equivalent assumptions is that v and δ in the above formula are constants.

Survivability hypothesis: It is assumed that the time scale of the virus is much smaller than the life cycle of the individual, thus ignoring the birth and natural death of the individual.

Under the above assumptions, by ignoring the degree correlation between different nodes, the response equation of density $\rho(t)$ of infected individuals can be obtained,

$$\frac{\partial \rho(t)}{\partial t} = -\rho(t) + \lambda \langle k \rangle \rho(t)[1 - \rho(t)].$$

In the above formula, the first consideration on the right side of the equal sign is to restore the infected node to health at a unit rate. The second item on the right side of the equal sign indicates the average density of the new infected node generated by a single infected node, which is proportional to the effective transmission rate λ, the degree of the node (assumed here to be equal to the average degree $\langle k \rangle$ of the network), and the probability $(1 - \rho(t))$ of connecting to the healthy node. Because we are concerned about the infection at $\rho(t) \prec 1$, other higher-order correction terms are omitted in the formulas.

If the right end of the upper formula is equal to zero, the stable density ρ of the infected individual can be obtained as follows,

Fig. 8.1 SIS model phase
diagram of uniform network

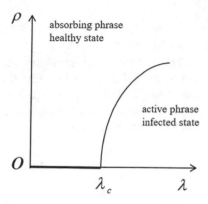

where the epidemic threshold is

$$\lambda_c = \frac{1}{\langle k \rangle}.$$

This shows that there is a limited positive transmission threshold λ in a uniform network. The infected individuals can spread the virus and make the total number of infected individuals in the whole network stabilize at a certain equilibrium state. At this time, the network is in an active phase. If the effective transmission rate is lower than this threshold, the number of infected individuals decreases exponentially and cannot spread widely. The complexes are in the absorbing phase at this time. Therefore, in a uniform network, there exists a positive critical value λ_c, which clearly separates the active phase from the absorption phase (see Fig. 8.1).

(2) Propagation Threshold of Scale-Free Networks

Now let us put aside the assumption of network uniformity and consider the propagation threshold of a typical non-uniform network, scale-free network. Relative density $\rho_k(t)$ is defined as the probability that a node with a degree of K will be infected. Its mean-field equation is

$$\frac{\partial \rho_k(t)}{\partial t} = -\rho_k(t) + \lambda k[1 - \rho_k(t)]\Theta(\rho_k(t)).$$

The unit recovery rate is also considered and the higher-order term $(\rho_k(t) \prec 1)$ is ignored. $\Theta(\rho_k(t))$ denotes the probability that any given edge is connected to an infected node. Note that the steady-state value of $\rho_k(t)$ is ρ_k. If the right end of the upper form is zero, it can be found that,

$$\rho_k = \frac{k\lambda\Theta(\lambda)}{1 + k\lambda\Theta(\lambda)}.$$

This indicates that the higher the degree of nodes, the higher the probability of infection. The non-uniformity of the network must be taken into account when calculating Θ. For uncorrelated scale-free networks, where the degrees of different nodes are uncorrelated, the probability of a node with a given edge directivity of S can be expressed as $sP(s)/\langle k \rangle$,

$$\Theta(\lambda) = \frac{1}{\langle k \rangle} \sum_k k P(k) \rho_k$$

If the above two formulas are combined, ρ_k and $\Theta(\lambda)$ can be approximately obtained for any scale-free distribution with sufficient small Θ. The propagating critical value λ_c must satisfy the condition that a nonzero solution of Θ can be obtained when $\lambda > \lambda_c$. From the above two formulas, we can get that,

$$\Theta = \frac{1}{\langle k \rangle} \sum_k k P(k) \frac{\lambda k \Theta}{1 + \lambda k \Theta}.$$

There is a trivial solution $\Theta = 0$. If the equation is to have a non-trivial solution $\Theta \neq 0$, the following conditions need to be satisfied,

$$\frac{d}{d\Theta} \left(\frac{1}{\langle k \rangle} \sum_k k P(k) \frac{\lambda k \Theta}{1 + \lambda k \Theta} \right) \Bigg|_{\Theta=0} \geq 1.$$

That is,

$$\sum_k \frac{k P(k) \lambda k}{\langle k \rangle} = \frac{\langle k^2 \rangle}{\langle k \rangle} \lambda \geq 1.$$

Thus, the propagation threshold λ_c of scale-free networks is obtained as follows,

$$\lambda_c = \frac{\langle k^2 \rangle}{\langle k \rangle}.$$

For scale-free networks with power law exponent $2 < \gamma \leq 3$, when the network size $N \to \infty$, $\langle k \rangle^2 \to \infty$, thus $\lambda_c \to 0$.

(3) **Propagation Critical Value of Associated Networks**

The critical value analysis described above is for irrelevant networks. In an unrelated network, the degree of any node is independent of that of its neighbors. However, it is found that relevance is an important feature of many complex networks, including the Internet.

Boguna and Pastor-Satorras studied the propagation threshold characteristics of correlated network [4]. Conditional probability $P(k'|k)$ is used to express the probability that a node with degree k is connected to a node with degree k'. If this probability is independent of k, it degenerates to the case where the node connection is unrelated. It is assumed that the network degree distribution $P(k)$ and conditional probability $P(k'|k)$ satisfy the normalization and equilibrium conditions,

$$\sum_k P(k) = \sum_{k'} P(k'|k) = 1,$$

$$kP(k'|k)P(k) = k'P(k|k')P(k') \equiv \langle k \rangle P(k, k').$$

The symmetric function $(2 - \delta_{kk'})P(k, k')$ denotes the joint probability of two nodes with degrees k and k', respectively. Define matrix $C_{kk'} = \{kP(k'|k)\}$. It can be found that the propagation threshold of the associated network is

$$\lambda_c = \frac{1}{\Lambda_m},$$

where Λ_m is the largest eigenvalue of matrix $C_{kk'}$. For the unrelated network, $C_{kk'}$ has only a unique eigenvalue $\Lambda_m = \langle k \rangle^2 / \langle k \rangle$, so the conclusion of the critical value of the unrelated scale-free network introduced earlier is reobtained. It can be seen that the critical value of propagation rate of complex networks is determined by the eigenvalue of matrix $C_{kk'}$ when considering the degree of network correlation.

Moreno et al., based on the study of SIR model, pointed out that [5], the epidemic incidence in the associated network is smaller than that in the non-associated network under the same degree distribution. Although the nature of the propagation threshold does not change with the emergence of relevance, the propagation declaration period is longer in the affiliated network than in the non-affiliated network. In addition, in the limited associated networks, the critical value of transmission rate is larger than that of non-associated networks, which indicates that the associated networks show stronger robustness against virus transmission than non-associated networks.

(4) **Research on the Broader Critical Value of Complex Network Propagation**

The degree distribution function $P(k) \propto k^{-3}$ of BA scale-free network. Now we consider a class of scale-free networks with variable power law exponent γ, called generalized SF network [6], whose degree distribution is represented by the following standard functions,

$$P(k) = (1 + \mu)m^{1+\mu}k^{-2-\mu} = (\gamma - 1)m^{\gamma-1}k^{-\gamma},$$

where m is the minimum degree of nodes in the network, $\gamma = 2 + \mu$. When $0 < \mu < 1$ (i.e., $2 < \gamma < 3$), $\lambda_c = 0$, and when $\mu > 1$ (i.e., $\gamma > 3$),

$$\lambda_c = \frac{\mu - 1}{m\mu}.$$

In particular, when $\mu > 2$ (i.e. $\gamma > 4$), the propagation threshold of scale-free networks is almost the same as that of exponential networks (i.e., uniform networks, such as small-world networks, random graphs).

Volchenkov et al. restudied the influence of power law exponent λ_c of scale-free networks on propagation threshold λ_c from the mechanism of evolutionary selection of biological networks [7]. They think it is always zero when $\lambda_c > 1$. They further pointed out that different scale-free networks do not have a common and effective immunization strategy for different priority selection principles, although they have the same power law index.

Based on the fact that the degree distribution of many practical complex networks is a mixture of power law distribution and exponential distribution [8], Liu et al. proposed that the priority connection probability Π_i obeys the corresponding combination,

$$\Pi_i \propto (1 - p)k_i + p.$$

When $0 \le p \le 1$, a hybrid network between uniform and non-uniform can be obtained, including scale-free networks and random graphs. It is found that for such hybrid networks, no matter how high the propagation rate λ is, a considerable part of the nodes will not be infected. This can also be understood as the existence of a nonzero propagation threshold λ_c in hybrid networks.

Complex networks with household structures have also begun to attract attention [9]. When separating the infection inside and outside the family, if the recovery rate of each family individual is faster than that of the infection inside the family, then there is no disease epidemic in each family, but the disease is easy to spread in the whole scale-free network.

8.1.2 Propagation Dynamics of Complex Networks

The critical value theory only considers the final steady state of propagation, while the analysis of dynamic behavior such as oscillation in the process of propagation needs to be considered in a new way. In this section, we mainly introduce some achievements in the study of the dynamics of the propagation dynamic equation of NW small-world networks. It is found that dynamic phenomena such as bifurcation, chaos, and general unstable oscillation can exist in the propagation model of small-world networks. The existence of bifurcations in scale-free networks also implies the oscillation behavior in the propagation process. Barthelemy et al. [10] analyzed the time evolution process of propagation outbreak in complex networks with highly non-uniform distribution. The result found that the virus propagation in non-uniform networks has a hierarchical dynamic behavior: the virus infects the moderately large nodes in the network at first and then invades the nodes in the network layer by layer to gradually reduce the moderately large nodes. Therefore, it is obviously very important to study the dynamic behavior of propagation dynamics in this chain propagation process.

(1) Linear Propagation Equation of d-dimensional NW Small-World Network

Viruses, disasters, fires, and congestion in communication networks can all be used to describe the propagation phenomena in networks. Newman and Watts first described the propagation process on a d-dimensional NW small-world network based on the NW small-world network model [11]. Moukarzel made a more specific analysis [12]. Here, Moukarzel's idea is used to introduce the propagation equation of this d-dimensional small-world network.

Let us assume that from the initial infected node A, the virus starts to spread at constant speed $v = 1$. The density of NW small-world network shortcut endpoint is $\rho = 2p$, where p is the probability parameter of adding new shortcut in small-world network model. It may be assumed that the propagation process is continuous. Therefore, the infection amount $V(t)$ of nodes in the network is a sphere $\Gamma_d t^{d-1}$ with radius t starting from A, where Γ_d is the hypersphere constant in the d-dimensional small-world network. The probability that the source of infection encounters the shortcut endpoint in the course of transmission is ρ, and the new infection sphere is $\rho \Gamma_d t^{d-1}$. Therefore, the average total amount of infection $V(t)$ is obtained from the following integral equation,

$$V(t) = \Gamma_d \int_0^t \tau^{d-1}[1 + 2pV(t - \tau)]d\tau.$$

After scaling and differentiating the formulas, the following linear propagation equations can be obtained,

$$\frac{\partial^d V(t)}{\partial t^d} = 1 + V(t).$$

Obviously, the solution of this equation is

$$V(t) = \sum_{k=1}^{\infty} \frac{t^{dk}}{(dk)!}, \quad d = 1, 2, 3, \ldots$$

is dispersed with the increase of time t.

(2) Fractal, Chaos, and Bifurcation of Small-World Network Propagation Dynamics Equation

Yang believes that in the infection volume $V(t)$ of NW small-world network, there is a time lag δ [13, 14] for the newly triggered virus infection or fire occurrence due to the waiting time in reality. Therefore, the corresponding linear time-delay propagation equation is

$$\frac{\partial^d V(t)}{\partial t^d} = 1 + V(t - \delta).$$

The corresponding solution is

$$V(t) = \sum_{k=1}^{t/\delta} \frac{(t - k\delta)^{dk}}{(dk)!}, \quad d = 1, 2, 3, \ldots$$

After calculating the fractal dimension D of the solution of the equation, where

$$D = \frac{d \ln(V(r))}{d \ln r},$$

it is found that δ determines the fractal dimension of NW small-world network.

Furthermore, the nonlinear friction in the process of spreading and spreading information flow in viruses, fires, and Internet and communication networks, such as competition among populations, congestion in networks, and restrictions of other media resources, will have an important impact on the process of transmission. Therefore, not only the time delay but also the nonlinear friction term $-\Gamma_d \int_0^t [\mu V^2(t - \tau - \delta)]\tau^{d-1} d\tau$ should be taken into account. There is,

$$V(t) = \Gamma_d \int_0^t \tau^{d-1}\left[1 + \xi^{-d} V(t - \tau - \delta) - \mu V^2(t - \tau - \delta)\right]d\tau.$$

After scaling substitution and differential order d, the following nonlinear propagation equations are obtained,

$$\frac{d^d V(t)}{dt^d} = \xi^d + V(t - \delta) - \mu\xi^d V^2(t - \delta),$$

where $\xi = \frac{1}{(2pkd)^{1/d}}$.

If the equation is written in discrete form, let $\delta = 1$, then the propagation equation of one-dimensional ($d = 1$) discrete small-world networks is

$$V_{n+1} = \xi + 2V_n - \mu\xi V_n^2.$$

When

$$\xi \geq \xi^* = \sqrt{\frac{1.401}{\mu}},$$

chaos appears in the system. When

$$\xi \leq \xi_0 = \sqrt{\frac{0.75}{\mu}},$$

the system tends to be a stable fixed point; when $\xi_0 < \xi < \xi^*$, the propagation process of period doubling bifurcation occurs [15].

For the nonlinear propagation equation, only one-dimensional case of $d = 1$ is considered, that is

$$\frac{\mathrm{d}V(t)}{\mathrm{d}t} = \xi + V(t - \delta) - \mu \xi V^2(t - \delta).$$

With μ as the bifurcation parameter, if $\delta < \pi/2$, then the Hopf bifurcation appears at

$$\mu^* = \frac{\pi^2 - 4\delta^2}{16\delta^2 \xi^2}.$$

(3) **Generalized Propagation Dynamic Equation and Bifurcation of Small-World Network**

Previous NW small-world network propagation equations are obtained through a series of scaling and time transformations, which are related to the probability parameter p of the NW small-world network model. Therefore, the evolution of network structure caused by the change of p cannot be revealed in the propagation equation by the scaling transformation. In addition, the nearest neighbor network at $p = 0$ is a special case of NW small-world network. However, the above equation cannot contain this particular network structure because $p = 0$ makes $\xi = \infty$, the equation is meaningless at this point. Most importantly, the term of nonlinear friction defined by Yang is fixed [16], that is, no matter how many shortcuts exist in small-world networks, nonlinear friction exists and remains unchanged. Intuitively, this friction should vary with the number of shortcuts that different p brings. Therefore, we define a new friction term $-\Gamma_d \int_0^t [\mu(1 + 2p)V^2(t - \tau - \delta)]\tau^{d-1}\mathrm{d}\tau$ and replace it with the equation. Still only $d = 1$ is considered, and the second equation is generalized as follows:

$$V(t) = \Gamma_1 \int_0^t [1 + 2pV(t - \tau - \delta) - \mu(1 + 2p)V^2(t - \tau - \delta)]\mathrm{d}\tau.$$

Let us set $\Gamma_1 = 1$. Thus, without any scale transformation and directly differentiating the above equation, a more generalized nonlinear propagation equation of the NW small-world network is obtained,

$$\frac{\partial V(t)}{\partial t} = 1 + 2pV(t - \delta) - \mu(1 + 2p)V^2(t - \delta).$$

The following Hopf bifurcation theorem reveals the effect of the small-world network model parameter p on Hopf bifurcation.

Theorem 4.1 *If $\delta < \pi/(4p)$, then when the bifurcation parameter μ passes*

$$\mu_0 = \frac{\pi^2 - 16\delta^2 p^2}{16\delta^2(1+2p)},$$

the Hopf bifurcation appears in the propagation equation.

When $p = 0.01$, $\mu_0(p = 0.01) = 1.5318$, order $\delta = \pi/4$. When $\mu = 1.2 < \mu_0$, the small-world network tends to be a stable fixed point; when $\mu = 2.1 > \mu_0$, the stable fixed point in the original network is no longer stable, and Hopf bifurcation appears, as shown in the figure, where $\mu = 1.2$ (solid line), $\mu = 1.6$ (dotted line), and $\mu = 2.1$ (dashed line).

By defining the following variables, the direction, stability, and periodicity of the corresponding bifurcation periodic solutions can be known by calculation,

$$\overline{B} = \frac{1}{1 - 2\sqrt{p^2 + \mu(1+2p)}\delta e^{iw_0\delta}},$$

$$C_1(0) = \frac{i}{2w_0}\left(g_{20}g_{11} - 2|g_{11}|^2 - \frac{1}{3}|g_{02}|^2\right) + \frac{g_{21}}{2},$$

$$g_{20} = 2\mu(1+2p)\overline{B}, \quad g_{11} = -2\mu(1+2p)\overline{B}, \quad g_{02} = 2\mu(1+2p)\overline{B},$$

$$g_{21} = -2\mu(1+2p)\overline{B}i$$

$$\times \left[-2\frac{g_{11} + \overline{g_{11}} + 2\mu(1+2p)}{2\sqrt{p^2 + \mu(1+2p)}} + \frac{-g_{20} - \overline{g}_{02} + 2\mu(1+2p)}{2iw_0 + 2\sqrt{p^2 + \mu(1+2p)}}\right],$$

$$a'(0) = \frac{8(1+2p)\delta}{4+\pi^2}, \quad w'(0) = \frac{16(1+2p)\delta}{\pi(4+\pi^2)},$$

$$\mu_2 = -\frac{\mathrm{Re}C_1(0)}{a'(0)}, \quad \beta_2 = 2\mathrm{Re}C_1(0), \quad \tau_2 = \frac{-\mathrm{Im}C_1(0) + \mu_2 w'(0)_-}{w_0}.$$

It is found that the dependence of bifurcation parameters on small-world network model parameters p is quite different under different delays δ.

(4) **Bifurcation and Oscillation of Infectious Dynamics Equations in Complex Networks**

The mean-field equations of SIS and SIR models in epidemiology have been introduced to study the transmission threshold. Recently, the stability of fixed points and bifurcations in infectious processes described by differential or difference equations have also attracted much attention. The dynamic equation of SIR model can be expressed by the following differential equation (parameters α, β, γ, $\delta > 0$, variable $S(t)$, $I(t)$, $R(t)$ are greater than zero),

$$\begin{cases} \frac{\partial S}{\partial t} = -\alpha SI + \beta R, \\ \frac{\partial I}{\partial t} = \alpha SI - \gamma I, \\ \frac{\partial R}{\partial t} = \delta I - \beta R. \end{cases}$$

Ramani et al. generalized the above SIR model to the difference equation of SIRS model as follows [17],

$$\begin{cases} I_n = I_{n-1} \frac{p+R_n}{1+R_n/q}, \\ R_{n+1} = \frac{a+bR_n}{1+I_n}. \end{cases}$$

It is found that this SIRS process does not simply tend to a stable fixed point, but can produce oscillating infectious behavior under some relative healing/infection conditions. Hayashi et al. [18] also used SIR models of uniform and non-uniform networks to study the propagation of computer viruses. When there is no immune measure, there exists a stable equilibrium point in both uniform and non-uniform networks, and the S, I, and R states of each node in the network converge to the stable equilibrium point in the form of damped oscillation. When the immune strategy is applied, if the number of infected nodes is equal to zero in the equilibrium state, there will be two situations in the non-uniform network: either there is a stable equilibrium point (the number of infected nodes is zero at this point), and finally the computer virus will be completely eliminated; or there will be a stable equilibrium point (the number of infected nodes at this point is nonzero) and a saddle point, which will appear saddle-node bifurcation.

8.2   Online Crowd Behavior and Influence Model

The topological structure, user interaction, and user content of social network constitute three elements of social network. Topological structure can depict the influence of nodes from the macrolevel, and it is easy to obtain. The index of topological structure in complex networks is relatively mature. Therefore, it is a common practice to measure the influence of nodes by using topological structure. However, the edges in the network topology cannot describe the complex interaction between nodes. For example, in microblogging, there are not only following among users, but also forwarding and commenting relationships, and the frequency of these interactions is different. Two close friends and two friends who meet each other are "treated equally" in the network topology structure. User behavior and interactive information can well reflect the formation and change details of user influence. Therefore, some researchers use the advantages of both to measure influence. Social network is the basis of user interaction, and the content of user interaction is the basis of user activity. Despite the analysis, users in different fields have different influence in their respective fields, so some researchers use content to analyze influence. By synthesizing three possible dimensions to measure impact, we illustrate the relationship between various studies of user impact and the characteristics of social networks from different perspectives.

There are different research perspectives and typical methods to measure influence from three dimensions: topological structure, behavioral characteristics, and content characteristics. We will divide the related achievements into three categories from three dimensions for review and analysis.

8.2.1   Measurement of User Node Impact Based on Topological Structure

Researchers in the related fields of sociology first used network topology structure to measure the impact of nodes. Then, researchers in other fields also carried out research and improvement. This section will introduce from four perspectives: based on local attributes, based on global attributes, based on random walk, and based on community relations.

(1) Measurement Based on Local Attributes

The most common measure based on local attributes is degree centrality, which is defined as the number of neighbor nodes. Degree centrality response is the direct influence of the current node in the whole network. For example, users who have a large number of fans in microblogs may have greater influence, but it is obviously not advisable to consider only the degree of consideration without considering the location of nodes in the network. Li et al. [19] proposed a centrality index for complex networks which takes into account the degree information of nodes and other neighbor nodes. Researchers found that when the propagation rate is small in the network, degree centrality has a better effect on the propagation of nodes, and when the propagation rate reaches the critical value, the measurement effect of eigenvector centrality is better [20, 21]. In addition, Ide et al. [22] also gave a graceful explanation from the point of view of dynamics. Lei et al. [23] further expanded the node degree by accumulating the neighbor node degree of the current node, proposed the index of ExDegree, and analyzed the number of layers suitable for information dissemination under different transmission rates. Fowler et al. [24] proposed the three-degree influence principle that: the node can affect not only the neighbor node (once), but also the neighbor node (second degree) of the neighbor node, and even the neighbor node (third degree) of the neighbor node of the neighbor node. As long as the three-degree relationship is strong, it has the possibility of triggering behavior. If there are more than three degrees, the interaction between nodes will disappear. The above indicators of degree-based expansion coincide with the ideas of Fowler et al. The metrics based on degree centrality and its improvement are simple, intuitive, and low time complexity, which are suitable for large-scale networks. However, this kind of index only considers the influence of nodes in terms of the number of nodes that may affect other nodes, and does not consider the difference between the strength of other nodes, and does not consider the location of nodes in the whole network.

In social networks, the phenomenon of forming associations with close friends is very common in social networks. Local clustering coefficient is used to measure the degree of tightness between neighbor nodes. The local clustering coefficient (hereinafter referred to as clustering coefficient) is equal to the ratio of the number of edges between neighbor nodes of node v_i to the maximum number of edges that can be connected between neighbor nodes. The formulas for calculating clustering coefficients of undirected graphs are as follows,

$$C(v_i) = \frac{2|\{e_{jk}: v_j, v_k \in N_{v_i}, e_{jk} \in E\}|}{k_i(k_i - 1)},$$

where k_i is the sum of the number of connected edges of node v_i pointing to other nodes and the number of connected sides of other nodes pointing to v_i, because in the undirected graph, the edges do not distinguish direction, so divide by 2. The formulas for calculating clustering coefficients of directed graphs are as follows,

$$C(v_i) = \frac{|\{e_{jk}: v_j, v_k \in N_{v_i}, e_{jk} \in E\}|}{k_i(k_i - 1)}.$$

In an example of a network with four nodes, under the three middle structures, the clustering coefficients of blue nodes are 0, 1/3, and 1. Researchers measure the influence of nodes by combining the degree of nodes and neighbors and the clustering coefficients of nodes. The results show that the clustering coefficients of nodes do not promote or even suppress the influence of nodes. Centol [25] studied the relationship between behavior propagation and clustering coefficients of nodes in online social networks and found that when the clustering coefficients of social networks are higher, the propagation speed of node behavior tends to be faster. The influence of nodes is related to the clustering coefficient of nodes. Ugander et al. [26] studied the formation and evolution of friendship in Facebook and found that the influence of nodes depends on the number of connected subgraphs formed between neighbors rather than the number of neighbors. Cui et al. [27] studied the persistence of information dissemination from the perspective of persistence. It was found that when information dissemination lasted for a long time, the dissemination range of information on tree network was wider than that on lattice network, which meant that nodes with large clustering coefficients did not necessarily promote the dissemination of information, that is, high clustering nodes did not necessarily have great influence. The research work of Ugander [26] and Mislove [28] shows that the clustering coefficients of nodes have a negative effect on the influence of nodes. Several close friends with high clustering coefficients form a close circle of friends. Chen et al. [16] carried out experiments on Scientist Cooperation Network and Short Message Network to verify that clustering coefficients play a negative role in node acquisition of new neighbor nodes. Based on the degree of nodes and clustering coefficients, ClusterRank, a measurement model of node influence, was proposed. In the model, the higher the aggregation coefficient of nodes, the smaller the influence of nodes. Zhang et al. [29] studied the influence of others on the word use of nodes in academic networks. On this basis, the role-conformity model (RCM) model was proposed to measure the trend of nodes affected by other nodes. It was found that nodes with larger degree but smaller clustering coefficient were vulnerable to the influence of other nodes. Taking the relationship between neighbors as a relevant factor of influence, the model improves the accuracy, but the time complexity increases.

(2) **Measurement Based on Global Attributes**

Node impact measurement index based on global attributes mainly considers the global network information of the network in which the node is located. These indicators can better reflect the topological characteristics of the node, but the time complexity is high. Most of the indicators are not suitable for large-scale networks. Betweenness centrality is defined as the number of times that the shortest path between two nodes in the network passes through the current node [30]. Median centrality describes the frequency of information passing through the node when it propagates in the social network. The larger the index value is, the busier the node is in the network topology. If the nodes with large median are removed, network congestion will occur, which is not conducive to information dissemination. Close centrality [31] measures the speed at which a node reaches other nodes. The larger the index value, the more paths the current node reaches to another node and the shorter the path length. This index can measure the indirect influence of nodes on other nodes. Eigenvector centrality [32] is an important index to measure the global impact of nodes. The centrality of eigenvector not only considers the number of neighbor nodes, but also the importance of neighbor nodes. The influence of a single node is regarded as a linear combination of the influence of other nodes. Similar to feature vector centrality, Katz centrality [33] is also considered in consideration of the different importance of neighbor nodes. But the optimal weight coefficients of this method need a lot of experiments to obtain, so there are some limitations. Kitsak et al. [34] through empirical research on social networks, e-mail networks, etc., found that high median or hub nodes are not necessarily the most influential. The nodes are divided into different levels from the edge layer to the core layer by K-kernel decomposition. It is considered that the core node (the node with large K value) is the most influential node. By iteratively subtracting the nodes whose degree is less than or equal to K, the nodes are divided into three layers, among which the nodes whose degree is $K = 3$ belong to the core node, that is, the node with great influence. The nodes with K value of 1 belong to the edge layer and have less influence.

The proposal of K-kernel decomposition has brought great inspiration to researchers. In recent years, many scholars have improved some defects of K-kernel decomposition and further improved its accuracy and scope of application. Liu et al. [35] found that the influence of nodes with the same Ks value may also vary greatly and then gave the ranking results of the influence of different nodes with the same Ks value. Zeng et al. [36] proposed an improved K-kernel decomposition method considering the residual degree of K-kernel decomposition. Moreno et al. [37] found that K-kernel decomposition was not effective when they studied the spread of rumors on social networks. Borge Holthoefer et al. [38] studied the activity of nodes, and the effects of different active distribution and distribution on K-kernel decomposition were obvious. Liu et al. [39] analyzed the core structure obtained by K-kernel decomposition in different real networks and found that the existence of closely connected small groups in social networks resulted in the existence of "pseudo-core" in many social networks. Considering the low information entropy of local small groups, a measurement index of influence based on network connection entropy is proposed. H-index is a common index to evaluate the influence of scholars or journals. Lu et al. [40] found that H-index also has a good effect in measuring the influence of nodes in social networks. It skillfully proves the correlation among degree, H-index, and the number of cores. The ratio of measurement effect of H-index on the impact of nodes and the number of cores of nodes has been significantly improved.

(3) Measurement Based on Random Walk

PageRank [41], HITS [42] and LeaderRank [43] are typical methods to measure the impact of random walk. The PageRank value of nodes can be calculated by considering the connection between nodes as the link between Web pages, so ranking can measure the influence of nodes. Because of the existence of isolated points and disconnected subgraphs, the original PageRank has the disadvantage that the ranking results are not unique. Liu et al. [43] improved this and proposed LeaderRank algorithm. Adding a two-way connection node to the original network solves the problem of non-unique sorting. Li et al. [44] improved LeaderRank algorithm by weighting. HITS algorithm is a sort method that takes into account both the center and authority of nodes. The influence measurement method based on random walk uses neighbor nodes to describe the influence of nodes. Although it avoids noise, it ignores the nature of nodes themselves.

(4) Measurement Based on Community Structure

The classical theory of weak connection proposed by Granovetterh [45] and Krackhardt [46] shows that there are "weak connection" and "strong connection" in interpersonal relationship. Granovetterh defines the strength of individual connection by the length of cognitive time, the number of interactions, and the degree of intimacy. For example, two close friends and colleagues who exchange work ideas every day belong to the "strong connection" relationship, while students who only meet occasionally belong to the "weak connection." According to Granovetterh, weak connections are better than strong ones, because weak connections can act as "bridges" between different groups. From the perspective of social network topology, close friends form associations, weak links correspond to sparse links between associations, while strong links correspond to close links within associations. Shi et al. [47] studied that in the data of various forums, user participation behavior patterns are closely related to the community structure and information dissemination influence of the communities they participate in. Zhao et al. [48] used the community partition algorithm to partition the social network. According to the number of communities connected by nodes, the VC index of nodes was proposed to further distinguish the influence of nodes.

21 nodes in the figure are divided into four communities, 2 links four communities, and 5 is located within one community. The VC value of node 2 is 4, and that of node 5 is 1. Therefore, the influence ratio of node 2 to node 5 is also 5, but the influence of node 5 within a community is greater. Hu et al. [49] proposed the K-shell and community centrality (KSC) influence model, which defined the degree, compactness, median, and K-shell of nodes as the internal attributes of nodes, the community structure of nodes and their neighbors as the external attributes of nodes, and the internal and external attributes of nodes as the external attributes of nodes. The external influence of the node is defined by the degree of association and the size of the community in which the node is located.

Burt et al. [50] proposed "structural holes" as a classical sociological theory. Because of the existence of structural holes, some nodes acting as intermediaries can obtain higher network benefits than their neighbors, that is, the importance of these intermediaries is greater. Su et al. proposed the N-Burt model to find the most influential nodes in the network. The model considers the nature of the neighborhood structure of the node and reflects the centrality of the community in which the node is located and the "bridge" of connecting different communities in the measure of influence of the node. Han et al. [51] utilized that the structure hole property and ListNet's ranking learning method effectively fused a variety of metrics including network constraint coefficients and achieved good results in the ranking of key nodes. Lou et al. [52] research showed that on Twitter, 1% of the structural hole nodes can control 25% of the information dissemination. Yang et al. [53] divided users into three types of roles when researching information diffusion, opinion leaders, structural holes, and ordinary users, and analyzed the role played by users as different roles in information diffusion. The influence of communication is more than 10 times that of ordinary users, and the structural hole nodes play a bridging role between different groups. The addition of a small number of structural hole nodes to the information dissemination process can promote the rapid spread of information.

The node influence index based on community structure considers not only the neighbor nodes of the nodes, but also the community nature of the neighbor nodes. Its advantage is to reflect the influence between individuals and groups. However, for the social network whose measurement results depend on the community nature of the social network and the community partition algorithm, the measurement effect is not good for the social network whose community structure is not obvious. Using topological structure to measure node influence is the most basic method. This method has multi-disciplinary theoretical basis and has achieved good results from the macrolevel of the whole social network. Some of the measurement indicators are simple and easy to calculate and have greater advantages in large-scale networks. However, the relationship between virtual network nodes is very different from that between individuals in the real world. For example, in the real world, individuals have different attributes, while in the topology, each node is the same and there is no distinction. Two intimately connected friends and a casual friendship are both sides of the topology. Obviously, the social network topology structure can only represent the whole social network from the macrolevel, but cannot describe the formation and evolution law of nodes' influence on other nodes at the microlevel. Social network topology makes little use of the behavior of nodes themselves and the interaction behavior of nodes to other nodes. For example, the topology regards one user's attention to another user as one side. However, there are many factors such as forwarding, commenting, and interaction time among users in microblog. The user's influence on a fan who forwards comments actively every day is obviously different from the fan who only follows but never forwards comments. The strength metrics based on the network topology structure and the advantages and disadvantages can be seen in Table 8.1.

Table 8.1 Influence metrics based on network topology

Indicator/method	Advantage	Disadvantage
Betweenness	Nodes with high information load capability can be found	Not suitable for large-scale networks
Clustering coefficient	Considering the close relationship between neighbors	Nodes with great global influence cannot be found
ClusterRank	Combining the advantages of degree and clustering coefficient, the accuracy is further improved	Not suitable for tree network
Closeness	Indirect influence between nodes can be calculated	The computational complexity of the algorithm is too high when calculating the global influence of nodes
Degree	Simple and intuitive, easy to calculate	It can only reflect the local characteristics of nodes
ExDegree	Expanding the degree, the accuracy of the ratio is high	Neighbor node location is not considered
Eigenvector	Can reflect the importance of neighbor nodes	Simple linear superposition without considering structure
H-index	The effect ratio and Ks of influence measurement were improved significantly	Multiple nodes with the same H-index can differ greatly
HITS	It combines the advantages of node centrality and authority	Poor noise resistance
KSC	It combines the advantages of location centrality of nodes with the advantages of connecting communities	Not suitable for tree network structure
LeaderRank	It has higher accuracy than PageRank, unique ranking result, and better anti-noise ability	Not applicable to undirected networks
PageRank	High accuracy of global arrangement in large-scale networks	Ignoring the node's own properties, the ranking result is not unique

(5)   **Influence Maximization**

Influence maximization is to let an influence spread to as many members of the social network as possible in a certain period of time. In the formal analysis, social network is a graph $G(V, E, P)$. V is the node set, e is the edge set, and P is the probability set of all sides. A user is a node v, the relationship between the user and the user is edge e, each edge has a probability p, and the information will be propagated on the graph according to the probability of the edge. The problem of maximizing the influence can be divided into two kinds: one is to select k nodes as the seed set so that the number of nodes that the seed set can affect is the most; the other is to find the minimum node set that meets the conditions given the required influence.

Linear threshold model and independent cascade model are the most common algorithms to maximize the influence. The independent cascade model assumes that each edge $e \in E$, which is associated with the probability $P(e) \in [0, 1]$. For any node u and any of its output neighbors v, u is the node activated at discrete time i, then v has the probability of $P(\langle u, v \rangle)$ being activated at time stamp $i + 1$. In other words, before u is activated, whether u can activate v has nothing to do with the diffusion history, so the order of node activation does not affect the diffusion result. For such a model, the diffusion process of the seed set is as follows: generally speaking, each newly activated node can independently activate its neighboring nodes according to the probability of the edge. In the linear threshold model, each node v contains the activation threshold θ_v selected randomly and uniformly from the interval $[0, 1]$. In addition, the linear threshold model specifies that the sum of all the access edge weights is at most 1, and the influence of other access nodes on it is cumulative. When the influence exceeds the threshold, the node is activated.

In terms of influence maximization, from the perspective of network structure, user influence can be maximized, such as measured by collective influence. In Fig. 8.2, $CI = (k_i - 1) \sum_{j \in \partial \text{ball}(i,\, l)} (k_j - 1)$ and $\partial \text{ball}(i, l)$ is the influence range of dimension l (radius) defined by the influence ball boundary.

It is assumed that the social level or ability of social network users can be displayed in the social network structure, and the network location of users is related to social conditions. We take social influence as the standard to measure the social level or capacity of users. The greater the social influence of users, the greater the social capacity of users and the more important the status of users in the network. We give two measurement methods.

1.   **Measure Method Based on Degree**

If a node i is connected to a larger node, that is, the larger the degree k_i of the node is, then we approximately think that the potential social influence of the node is greater. We make $p_i = 1/k_i$, which represents the reciprocal of the degree of node i. Through the formula $-(1 - p_i) \log_2 p_i$, the potential social influence of this node i is measured. At the same time, we use $\text{ball}(i, l)$ to represent the set of points in the range with node i as the center and the size of l (radius) (excluding node i itself), $n = |\text{ball}(i, l)|$, indicating the number of nodes in the set. $p_j = 1/k_j$ (k_j is the degree of node j, j $\text{ball}(i, l)$).

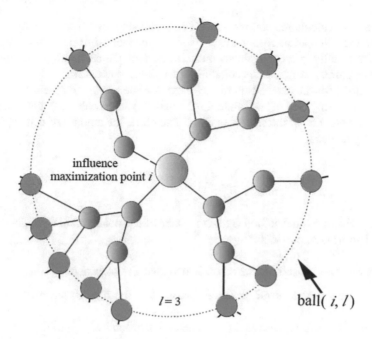

Fig. 8.2 Maximum influence

Then the social influence of node i is expressed as follows,

$$\text{SI}(i) = -\frac{\alpha}{n} \sum_{j \in \text{Ball}(i,l)} (1 - p_j) \log_2 p_j - \beta(1 - p_i) \log_2 p_i, \quad \alpha + \beta = 1.$$

This formula measures the social influence of users with different values of α and β, respectively, and seeks the best combination of self and external factors influencing social influence of users.

Previous studies have shown that the location determinant of nodes in network organization is not its own attribute (such as degree), but its importance or propagation influence.

We can further consider the importance of user location: in the social network, the average path length L of the network is defined as the average value of the distance between any two nodes, i.e., $L = \frac{2}{N(N+1)} \sum_{i \geq j} d_{ij}$, and the average path change value before and after removing the node I is expressed as $\delta = |\Delta L|$, which can reflect the importance of the node or user's position in the network structure from the perspective of the change of the average path length. Therefore, we redefine the user's social influence as,

$$\text{SI}(i) = -\left[\frac{\alpha}{n} \sum_{j \in \text{Ball}(i,l)} (1 - p_j) \log_2 p_j + \beta(1 - p_i) \log_2 p_i \right] * \delta(\alpha + \beta = 1).$$

In fact, the calculation complexity of δ in social network is relatively high, which leads to the calculation complexity of social influence of users in this chapter is higher, especially in large-scale social network data, the calculation amount is too high. Therefore, this method is suitable for small-scale networks.

Therefore, for the calculation of δ, we can use the average path change value in the network composed of the points in the ball(i, l) set to calculate and control the calculation amount by the size of radius l. Therefore, the amount of calculation can be greatly reduced,

$$L = \frac{2}{N(N+1)} \sum_{i \geq j} d_{ij} \, (i, j \in \text{ball}(i, l)), \quad \delta = |\Delta L|$$

The problem of this method is that the value of δ is not a global value, but a value of local position importance.

2. **Measurement Method Based on Integrated Environment**

We consider the node itself, external factors, and the importance of the network location.

Let $p_i = \mathrm{e}^{-\delta k_i}$ (δ represents the relationship between k and p_i of the node, which is determined by the average degree path of the node and the whole network). Then the user social influence can be defined as,

$$\text{SI}(i) = -\frac{\alpha}{n} \sum_{j \in \text{ball}(i, l)} (1 - p_j) \log_2 p_j - \beta(1 - p_i) \log_2 p_i, \quad \alpha + \beta = 1.$$

Similarly, in order to reduce the computational complexity, the calculation of δ in the formula can be calculated by the average path change value in the network composed of points in the ball(i, l) set.

Therefore, the importance of node location can be taken into account to measure the social influence of users.

The above two measurement methods can be used to discover the influencers, leaders, experts, and other roles in social network users from the perspective of network structure and to identify social users based on network structure.

Of course, a measurement method is only considered from the perspective of network structure. In fact, the social influence of users is closely related to the content generated by users and the process of content dissemination. The social influence of a user should be measured by considering its network structure, social content, production, and popularity.

In social networks, the measurement of influence nodes and the technology of maximizing influence have become hot issues in the social field. Among them, the research on maximizing influence has important practical significance for controlling disease outbreak, improving advertising effect, optimizing information dissemination effect, and mining influential individuals of social networks. At present, the measurement of influence nodes mainly locates the nodes which play an important role in information dissemination through the network topology such as neighborhood, location, path, and so on. For example, degree center, intermediate center, proximity center, Katz center, k-core center, and other classical center measurement methods have been widely used to measure network influence nodes. Among them, there are many heuristic methods based on network centrality, such as maximum method, shortest average distance method, PageRank method, and so on, many greedy methods based on sub-modules, such as greedy algorithm, CELF algorithm, and later NewGreedy and CELF++, and many methods based on objective function optimization, such as simulated annealing algorithm.

(6) **Popularity Prediction**

The purpose of popularity prediction is to predict the popularity of online goods in the future by using the interactive data of goods and users in the online system, such as color, material, silhouette, style, matching, and the concept of life pursuit of mainstream consumers. Among them, the research on the popularity prediction of online social media has great significance both in theory and in reality. On the theoretical level, popularity prediction can help people understand which factors may lead to the popularity of commodities, the dynamic evolution of commodity popularity, and how microbehaviors converge to macrocommodity popularity. In reality, the prediction of commodity popularity can help online users filter out useless information more effectively, online service providers better organize and manage their platforms, and provide strategic guidance for advertisers.

At present, social network has become an important platform for people to acquire and share information. The documents generated by users in social network are advancing with the times. Among them, topics of interest to users also evolve over time. Topic has always been an important research direction in the field of social network. Traditional topic popularity is mostly aimed at explicit topics, so it is difficult to explain the macroprocess of topic communication and evolution. So we can use topic model to get implicit topics from document collection. In order to quantitatively measure or predict the popularity or popularity of topics, we can use dynamic topic model (DTM) to obtain the topic distribution of each user in each time slice of social network. On this basis, a factor graph model is established to predict the possible topic distribution in the future and then calculate the popularity of topics. The dynamic topic model considers four factors: user interest, neighbor influence, topic autocorrelation, and cross-topic correlation. Specifically, the user's interest in a topic at time t is affected by the user's interest in the past and also by the user's neighbor nodes. Topic relevance also affects the user's interest in a topic. The model uses iterative confidence propagation algorithm to calculate the normalization factor of the model and gradient descent method to estimate the parameters of the model. After learning the parameters of the model, we can predict each user's interest in the topic in the future and then calculate the popularity of the topic. At the same time, DTM model is dynamic, which means that topics in social networks will be updated with time.

In fact, DTM model is a model proposed based on LDA. DTM added time series information to train time series-related model, so as to analyze the change of theme in time series. The traditional LDA theme model considers that the theme corresponding to each word in the document can be sampled from a series of themes interchangeably, but in the real world, the time sequence of the document affects the theme in the theme collection. DTM assumes that the text can be divided into multiple sets in chronological order, such as by year, so the theme of time slice t is evolved from the theme of time slice $t - 1$.

8.2.2 Impact Measurement Based on Content and Behavior Characteristics

(1) Measurement Based on Information Content

In social network, there are not only links among users, but also text content of information released by users. Information content is the carrier of influence dissemination. Combining with information content of users, it helps to analyze the mechanism behind influence promoting information dissemination. Matumura et al. [54] proposed an influence diffusion model based on the content of response information. If user y responds to the content of user x, the degree of y affected by x is determined by the similarity between the content of user x and that of user x. The calculation formula is as follows,

$$i_{x,y} = |w_x \cap w_y|/|w_y|$$

Among them, $i_{x,y}$ represents the influence of user x on user y, w_x is the collection of information items (words) of information content published by user x, and w_y is the collection of information items of user y's reply content to the information published by user x. If user z continues to respond to user y's content, the indirect influence of user x on user z is defined as,

$$i_{x,z} = (|w_x \cap w_y \cap w_z|/|w_z|)i_{x,y}$$

Matumura et al. measure the direct and indirect influence of users by the similarity of reply content between users, which has a better measurement effect on the local influence of users. However, the IDM model does not analyze the structure of user response, which has some defects.

Duan et al. [55] determined the impact degree of the respondent on the respondent according to the tendentiousness or emotional transformation degree of the respondent in the forum. A TTRank model based on tendentiousness transformation was proposed to measure the impact of the user on others and the degree of the influence of others in the process of posting. Finally, the ranking results of the user influence were obtained. Fan et al. [56] analyzed the possible problems in IDM model, such as reply chain interruption, content delivery interruption and false reply, improved IDM model, proposed impact diffusion probability model (IDPM), and applied the model to the problem of opinion Leaders' discovery. According to the similarity of information content and the change of emotional attitude of content, the influence among users is measured. Content-based impact measurement model can analyze the specific forms of users' influence on others, but it requires high content of data text, which is not conducive to the measurement of users' influence in large-scale social networks.

In large-scale social networks, the novelty of user-generated information content is one of the factors affecting the popularity of user-generated information, and the popularity time and scope of user-published information are also the basis for measuring user influence. Song et al. [57] considered the novelty and topology of user's information content when analyzing the influence of users in blog, extended PageRank algorithm, and proposed InfluenceRank algorithm to discover the influential users in blog social network. Agarwal et al. [58] used the length, novelty, and number of references and replies of blog posts in blog data to measure users' influence. Bakshy et al. [59] used the tree structure of information diffusion on Twitter to measure the influence of users and predicted the influence of users. Research has found that many popular topics are initiated by influential users. However, Bakshy and others failed to accurately measure the impact of users. When Peng et al. [60] studied the ultimate popularity of microblog information, they found that the topological structure of the early forwarders of information dissemination was closely related to the ultimate popularity. If the information sent by users is widely disseminated in the early stage, it will be conducive to the widespread dissemination of information in social networks.

In social networks, users have their own topics of interest. Researchers have found that users have different influence on different topics. Measuring the influence of users on different topics can give a more detailed description of users' influence from different perspectives. Researchers use statistical machine learning to measure users' influence on different topics. Assuming that the user's influence on another user is an implicit variable, the machine learning method learns the implicit variable iteratively through Gibbs sampling or EM algorithm. The latent Dirichlet allocation (LDA) model proposed by Blei et al. [61] can simulate the process of document generation, so that the LDA model can be used to obtain the distribution of information content sent by users on different topics. Dietz et al. [62] synthetically utilized the text content and user's topology of the information sent by the user. Assuming that there are two reasons why the user produces the text content, the user's own creation produces or is influenced by others. The source of the text content and its corresponding probability are calculated iteratively through Gibbs sampling, so as to obtain the influence of the user on different topics. Tang et al. [63] used LDA to obtain the topic distribution of users on different topics in large-scale heterogeneous networks and proposed a topical affinity propagation (TAP) model to measure users on different topics in combination with topic similarity or strength among users. The intensity of the influence is solved by Gibbs sampling, and then the PageRank is expanded to find representative users on different topics, that is, users with great influence on different topics. Li et al. [64] used a unified probabilistic model and topic-level opinion influence model (TOIM) to integrate topic factors, user views, and node influences into the same model. The model has two learning stages: the first stage combines topics factors and user influence to generate influence relationships among users on different topics. In the second stage, the TOIM model is used to construct historical viewpoints and neighbors' opinions by using user history information and interaction relationships, thereby further predicting users' views on a certain topic. Guo et al. [65] improved LDM model by using social network, put forward social-relational topic model (SRTM) topic model, and applied it to measure user's topic influence. When learning user topic distribution, the model considers not only the content of information sent by users themselves, but also the content of topics sent by their neighbors. The model uses Gibbs sampling algorithm to solve the parameters and uses logistic regression to determine whether the user's attention behavior is caused by the topic influence of the user concerned or by the global influence of the user concerned, so as to obtain the topic influence and global influence of the user. Weng et al. [66] proposed TwitterRank to measure users' influence on different topics based on PageRank, which synthesized the similarity of users' interest and network topology on Twitter. Cui et al. [67] proposed a hybrid factor non-negative matrix factorization (HF-NMF) model based on interactive information between users on social networks. HF-NMF model can measure and predict the influence of users on different information items. Tsur et al. [68] combined content and time factors with network topology and used linear regression to predict the popularity of tags on Twitter in a certain period of time.

Some scholars have also studied how to measure users' influence according to content when the network topology is unknown. Gerrish et al. [69] proposed a dynamic topic model when analyzing the impact of papers in data sets of papers without links and applied it to measure the impact of papers. Shaparenko et al. [70] used the probability model of machine learning to measure academic influence. Romero et al. [71] studied the popularity and influence of Twitter tags. Influences measurement methods and models based on information content can describe in more detail the specific forms that users show when influencing others. This kind of influence may lead to similarities and consistency between other people and users in information content and may also lead to changes in other people's emotional attitudes on a topic. However, this kind of method ignores the relatively stable influence formed during the long-term communication between users.

(2)   **Measurement Based on User Behavior**

The quantitative analysis of human behavior and the study of the temporal and spatial laws of human behavior have attracted researchers in the fields of complexity science, statistical physics, etc., and have generated great interest. The study of the temporal and spatial characteristics of human behavior on user behavior in social networks modeling, analysis, and propagation dynamics research play an important theoretical guiding role, which further provides a theoretical basis for measuring user influence through the behavior of users in social networks. The researchers conducted a detailed review of the progress of the analysis of the temporal and spatial characteristics of human behavior. Users in social networks publish content information and then disseminate the content through interactive behavior. By analyzing these behaviors, not only can the influence intensity of users be measured, it can also predict the speed and scope of user behavior on social networks. User behavior based on metrics is usually the influence between nodes and belongs to local influence. Local influence can be transformed into global influence by weighted average and random walk.

User behavior in social networks is recorded in the form of logs, which can measure the strength of influence among users and the spread range and speed of user information and behavior in social networks. Yang et al. [72] used exponential function or power law function to measure the number of times that other users mentioned certain information in a certain period of time when they analyzed the changing rule of user influence over time on Twitter. They proposed a linear influence model (LIM) to measure user influence and predict the spread range of user information in a short time. The model does not use the network topology among users. Tan et al. [73] synthetically considered data such as topological structure and user history behavior to measure the influence of neighbors on users, thus predicting user behavior. Trusov et al. [74] analyzed the relationship between user's activity and user's influence in social network and found that if the number of friends increased rapidly during user's active period, the user's influence would be greater. Goyal et al. [75] used machine learning to measure the strength of influence among users from the behavior of sharing pictures among users of Flickr Web site. The frequency of behavior propagation was used as the index to measure the influence of users, and the scope of behavior propagation was used as the index to measure the influence of behavior. Xiang et al. [76] used interaction behavior and topic similarity between users to measure their influence. Zaman et al. [77] used collaborative filtering algorithm to measure the influence of users' forwarding blog based on their forwarding behavior and content. Yang et al. [78] studied the user's forwarding behavior in Twitter and found that the user's forwarding behavior obeys power law distribution and is influenced by the user's influence, content, and time factors. Cha et al. [79] used attention behavior, forwarding behavior, and mentioned behavior to measure the user's attention influence, forwarding influence, and mentioned influence and analyzed the evolution law of user influence over time. It was found that the users concerned by a large number of users did not necessarily attract the attention of the user's forwarding behavior and mentioned behavior. Mao et al. [80] took into account the user's reading habits and forwarding preferences in microblog to measure the user's influence.

The above model starts from the user's behavior or measures the user's influence or the influence intensity between users. It establishes a connection between user's behavior and user's influence and achieves good results. However, this kind of model regards the behavior of a pair of users as irrelevant to that of other users, does not consider the complex relationship between users and their multiple connected users, and does not make use of or make full use of the topology of social networks. Therefore, it is one of the hot research trends of current researchers to improve the accuracy and efficiency of the quantitative user impact model by synthetically utilizing the user's behavior characteristics and the social network's topological structure.

Zhang et al. [81] studied the forwarding behavior of microblog users. It was found that the probability of users' forwarding messages was affected by the number of connected graphs formed between neighbors who had forwarded this message. The social influence locality was proposed to measure the influence of the structure of user's circle of friends on forwarding behavior.

Romero et al. [82] considered not only the degree of popularity and activity of users themselves, but also the passivity of their friends when measuring the influence of users. The stronger the passivity is, the less susceptible to the influence of others. Bao et al. [83] found in the analysis of the popularity of information published by users in Weibo: the final popularity of information is related to the topology characteristics of users who forward this information early, and there will be connections of interest in early forwarders. The ratio of the number to the total number of possible connections is defined as the link density, and the maximum distance from all forwarders to the information initiator is defined as the diffusion depth. The research shows that the lower the connection density of early forwarders, the higher the diffusion depth, and the higher the final popularity of information. The influence of users is closely related to the popularity of information released by users, and the more influential users are more likely to cause topics of greater popularity. The above model fully analyzed the influence between users and their associated groups, and combined with the user's behavior, the user's future behavior and information dissemination were predicted and achieved good results. However, this kind of model does not directly measure the influence of users.

Researchers have extended PageRank algorithm from different perspectives. Haveliwala et al. [84] comprehensively considered the user topic tendency and the sensitivity and novelty of publishing information. On the basis of PageRank, personalized PageRank algorithm was proposed to measure user influence. Tang et al. [85] proposed topical affinity propagation (TAP) as a user's propagation model on topics, and combined with PageRank, proposed a PageRank with topic-based impact (TPRI) model to measure users' global influence on different topics. Xiang et al. [86] studied the relationship between user authority and influence in scientist cooperative networks, proposed a linear influence model, and improved PageRank's user authority measurement by introducing prior knowledge. Weng et al. [82] proposed TwitterRank, believing that users have multiple topics of interest and have different influence on each topic. Considering the topic similarity and topological structure among users, PageRank was expanded to measure the impact of users on a single topic. Ding et al. [87] defined four kinds of relational networks according to users' reading behavior, forwarding behavior, replication behavior, and reply behavior in Twitter and proposed a topic hierarchy impact model (MultiRank) based on multi-relationship network. Based on the influence of the user on a single topic, the influence of the influence on the four networks and the network is used to calculate the user's comprehensive influence on multiple topics. Wang et al. [88] synthetically measured the influence of opinion leaders in Weibo from three aspects: the initial influence of users, the persistence of influence in the process of dissemination of Weibo messages, and attenuation index. When studying the influence of Web sites in Web networks, it is different from traditional research methods in utilizing the information flow and connection relationship between Web sites. Jin et al. [89] put forward a gated attention model with LSTM-CRF neural networks by analyzing the change of user's attention.

(3) Measurement Based on Time Factor

Time-related factors such as interaction time and behavior time play an important role in information dissemination in social networks. Taking time factor into account in the impact measurement model can effectively improve the accuracy of the model. Some characteristics of human behavior, such as paroxysmal and memory, are also applicable in social networks. Huang et al. [90] analyzed the probability of forwarding information among users in microblog and found that the time interval between one user forwarding another user's information and the last time of forwarding the user's information presented the characteristics of short-term burst, high frequency, and long-time silence and obeyed the power law distribution with different parameters. Bayesian learning method can be used to model and predict the probability of information forwarding among users. Li et al. [91] proposed a generalized uncorrelated regression with adaptive graph for unsupervised feature selection. Zhao et al. [92] used Bayesian model to measure the direct influence among users according to the self-similarity and interaction times of nodes in time domain, and semi-ring algebra to measure the indirect influence among users. By adding "false fans" to the original social network, the stability of the ranking results of user influence obtained by the model is verified. Pan et al. [93] used the change of user's status information to model the user's influence under unknown conditions, such as user's behavior, and analyzed the change of user's influence with time. Qin et al. [94] proposed a dynamic connection-based social group recommendation based on the node influence. Rodriguez et al. [95] inferred the potential network topology structure and the propagation rate of information among users by using the time delay of information diffusion based on historical information diffusion data. NetRate algorithm was proposed and solved by stochastic convex optimization.

In a word, the influence of users cannot be seen or touched. However, the changes that users make other users in social network through their own influence can be observed, which can be expressed as similarities in content and consistency in behavior, or changes in attitudes and concepts. The influence measurement method based on user information and behavior characteristics can better describe the formation and development of influence between users and users compared with the influence measurement method based on topology. In social networks, there are many forms of user behavior. Existing models mostly consider only one of these behaviors or separate the various behaviors, measuring user influence by simple weighting, without considering the interaction and mutual difference between these behaviors and user information content. If a user has a great influence on another user, it may be manifested by the number of interaction actions, the similarity of their behavior in time series, and the short time interval between their actions. However, there are few

Table 8.2   Advantages and disadvantages of impact measurement methods

Impact measurement method	Advantage	Disadvantage
Measurement based on network topology structure	The applicability of various social networks is strong, and the macrolevel has guiding significance and certain advantages in large-scale networks	Ignoring the diversity of user–user relationships
Measurement based on content characteristics	More specific measures of influence	High requirement for data preprocessing
Measurement based on behavior characteristics	The prediction accuracy is high and the influence propagation caused by user behavior is considered	Applicable only to specific social networks

intrinsic relationships among these behavioral characteristics, which can be reflected in existing user impact models. It remains to be studied to further clarify the relationship between the complex relationships among users' various behaviors and users' influence in social networks. We compare the advantages and disadvantages of three kinds of impact measurement methods, as shown in Table 8.2.

8.3   Online Social Networks User Role Identification

It is of great significance to identify the roles in social networks: (1) It is of great significance to identify the specific roles in social networks. For example, identifying the role of experts in technical forums can help us get the best answer to questions, identifying the role with great influence in social networks has an important role in supporting the diffusion of enterprise product information and the establishment of good reputation, identifying the role of opinion leaders in social networks plays a key role in guiding the correct direction of public opinion, and so on. The behavior of some specific types of network users, such as opinion leaders, important users, and network navy has a very important impact on the generation and dissemination of information in the network and has become an important force that can influence the development and trend of network public opinion. In particular, the role of opinion leaders in social networks plays an important role in the formation of public opinion. Under its guidance, some local opinions will evolve into public opinion, which directly affects the real society. Identifying the role of social network is of great practical significance not only for us to make full use of the network, but also for the monitoring and management of social network. (2) Identifying the role of social network has important research significance. Identifying and analyzing user roles provide a valuable perspective for us to study the dynamic characteristics of the network and are of great help to the in-depth study of social networks. For example, the user's behavior represents a role to some extent. When a new social relationship emerges in the social network, we can better understand the interaction between them according to the identification of the roles in the social network. By analyzing the roles in the social network, we can further understand the network topology structure and understand the role of the temporal evolution of the network on the formation and change of the user's roles. It will also make it easier for us to understand the process of information dissemination in the network. In addition, the change of roles in the community is directly related to the development and change of the community, and the change of roles can lead to the change of the community. Therefore, the identification of roles in social networks is of great help and significant for us to analyze and understand social networks, predict user behavior, and study the relationship and interaction process between users.

8.3.1 Uncertain Role Recognition Method

At present, the recognition of unclear roles is mainly realized by machine learning and mathematical analysis. Uncertain roles mean that there is no (or very little) background knowledge to define roles in advance, which will be identified mainly through social network structure, feature description, and text information generated during interaction. Among them, machine learning mainly takes into account network structure, interactive information, or both. Through unsupervised learning process, data (or nodes) are automatically divided into different roles, and user roles are finally recognized. For example, Laurent et al. [96] proposed an improved algorithm by using machine learning to replace decision tree stumps with small decision tree, which solved the problem of speaker role recognition in radio news programs.

Block models [97] and probabilistic models are usually used in mathematical analysis to analyze and identify [98], which is also the main method of unclear role recognition. The role identified here can be understood as the location of the network node. Just like the position of "manager-secretary" in a company, the position of "manager" can be identified according to the vocabulary in the e-mail used by the company. Then, according to the position of the recipient of the e-mail, it can be judged that this is a "secretary" or another "manager." In addition, non-explicit roles or nodes can be identified by the clustering process of graph structure or interactive information.

(1) Block Model Recognition Method

Block models are built on the basis of a social network and use predefined equivalence. Their purpose is to classify predefined equivalent 0/1 functional blocks. In the block model recognition method, the most important thing is to correctly define the equivalence of a relationship according to the problem being handled. At present, the equivalence of several relationships depends on the structural equivalence, regular equivalence, strong equivalence, and automorphism equivalence of the object of study. Structural equivalence generally refers to the grouping of members with similar interests, while regular equivalence and automorphism equivalence are more sociological concepts used to express social roles: people with the same role can only contact people who share the same role. Usually, a block model is a simple graph structure represented by a two-dimensional 0/1 matrix, which is related to some kind of relationship (such as colleague relationship, classmate relationship, and friend relationship). Block models based on related data can be used to explain the activities of complex networks. The complex social network is simplified into one and several simple graph structures by choosing corresponding equivalents through block modeling. The processing process is shown in the following figure. The two-dimensional relationship matrix shown in figure (b) can be obtained from the social network (a) composed of the original 10 persons, in which the relationship between the two is expressed in 1. The rearranged 0/1 matrix (c) is obtained by permutation of rows and columns. It can be seen from the graph that the matrix consists of four blocks: three 0 blocks and one block containing one. By using predefined criteria and related equivalent concepts, the block model (d) is obtained. Block model (d) can be expressed as the relationship $A \rightarrow B$ between locations A and B. For example, it can represent the role network (e) of teacher $A(T)$ and student $B(S)$.

Block model processing technology can be said to be one of the most frequently used technologies in social network analysis. On the basis of guaranteeing its mathematical theory, block model is summarized to analyze various types of network structures [99]. At the same time, block modeling is also one of the main methods to obtain the structural model of social network. Although block modeling is mainly aimed at network structure, it can also be used to deal with the attributes and multiple relationships of nodes. In the process of role recognition of block model, the location of nodes can be given by analysts or estimated in an unsupervised way. If the type of location is not predefined, the same cluster and the relationship between clusters need to be calculated at the same time. This is the location or role, which means that the role here is the estimated location.

In previous studies, regular equivalence and structural equivalence are commonly used to study networks. In REGE and CATEREGE algorithm [100], the relation of data (i.e., network) is regarded as input, and the function of set partition as output is regarded as the definition of regular equivalence. The degree of regular equivalence between node pairs is calculated, and block model is established to identify the hidden location. According to the equivalence of network structure, CONCOR algorithm divides the data into two blocks and constructs a hierarchical tree (tree graph) by iteration process [101]. Ultimately, each original node is associated with a position or role in the graph. In addition, there are some stochastic models that are mainly concentrated on structural equivalence [102, 103], and the results of the partitioning are obtained by characterizing the network nodes (e.g., social categories), which can be regarded as a stochastic block model (stochastic block models). Later, there are further analyses that extend these models to potential classes, except that these potential classes do not assume that cluster members are known, but rather estimate them from the data [104]. Compared with the previous methods, Handcock et al. [105] integrated the probability model of distance between individuals. The proposed LPCM model takes into account the transitivity of graphs, which shows that when people have common attributes (such as age, gender, race, and geographical location), they tend to be more interconnected. Wolfe et al. [106] further extended the previous random block model by allowing the use of multiple roles. In addition, Airoldi et al. [107] proposed a mixed membership stochastic block model (MMSB) for relational data and a general variational reasoning algorithm for fast approximation and inference. The previous latent random block model was improved to allow each object to belong to several different clusters at the same time. That is to say, a person can play several different potential roles at the same time. Later, Fu et al. [108] took into account the natural evolution of the network, that is, the role can change with time, and proposed a new dynamic mixed membership block model (DMMB).

In conclusion, in order to adapt to the traditional social network model, a mixed membership stochastic block model (MMSB) is proposed. Similar to the hybrid model clustering method, each node is associated with a membership vector, which involves different clusters and ultimately obtains the location or role of the node or individual.

(2) Probabilistic Model Recognition Method

When we analyze text data sets such as e-mails, blogs, academic papers, and so on, we may find that it is not enough to rely solely on the relationship structure to analyze or identify the roles in them. In this case, we need to use another main method of analyzing and identifying roles, namely probabilistic model method [98]. Probabilistic model method mainly deals with text data sets, usually using unsupervised hierarchical Bayesian model to achieve. Without considering the network relationship structure, the text content is associated with the edges in the graph. The implementation process mainly depends on the existing topic model, which first assumes a probabilistic generation model, then associates each text with multiple topics, and finally extracts topics from the text. At present, probabilistic topic model is usually a way of thinking, that is, text is regarded as a random mixture of several topics. In different models, there are usually different statistical assumptions and different methods to get model parameters.

In thematic model, a topic is usually defined as a polynomial distribution of a given word. Thematic model [109] summarizes a large number of texts with a smaller number of words, which are called "themes." For example, in the author-topic model [110], each author is associated with a polynomially distributed topic, and each topic is associated with a polynomially distributed word. In this generation model, each document is represented as a mixture of topics. For example, in latent Dirichlet allocation (LDA) method [61], by allowing document authors to determine the mixed weights of different topics, these methods are extended to author modeling. By learning the parameters of the model, we get a set of themes that appear in the corpus and their related documents and identify the authors and the themes they use. In the LDA model, the generation of document sets can be simulated as three processes. Firstly, for each document, the theme distribution of the document is extracted from the Dirichlet distribution. Then, according to the topic distribution, a single topic is selected for each word in the document. Finally, each word is sampled from the vocabulary polynomial distribution with a specific sampling topic. The hierarchical Bayesian model corresponding to the generation process is shown in the following figure. Among them, Φ denotes the topic distribution matrix and independently characterizes T topics in the existing symmetric Dirichlet (β) distribution through the polynomial distribution of V words. θ is a matrix of mixed weights of T-themes for a specific document, which independently characterizes each topic in the existing symmetric Dirichlet (α) distribution. For each word, Z means that the subject responsible for generating the word is derived from the distribution of the document, W is the word itself, and the subject distribution Φ corresponds to Z. Estimate the values of Φ and θ parameters to get the subject information about the participants and the weight information of these topics in each document.

On the basis of LDA research, Rosen-Zvi et al. [111] proposed an author model in which each author is associated with the distribution of a word rather than a topic, where x denotes the author of a given word and is randomly selected from the author set a_d. Each author corresponds to a word's probability distribution Φ, which is also derived from the existing symmetric Dirichlet (β) distribution. By estimating the value of Φ, we can get the information about the author's interest, at the same time, we can get the information about similar authors and authors who are similar to the theme of the document. Based on these two models, the author-topic model (ATM) is proposed. Similar to the author model, x represents the author of a given word and is also randomly selected from the author set a_d. Each author corresponds to a topic's probability distribution θ, which is also derived from the existing symmetric Dirichlet (α) distribution. According to the mixed weights of the selected authors, a topic z is selected and a word W and Φ distribution are obtained from the existing symmetric Dirichlet (β) distribution.

Author-topic model can be said to be a combination of LDA topic model and author model. Just as every document in LDA topic model has a unique author, each author in the author model has a unique topic. By estimating the values of parameters Φ and θ, we can get the topic information that the author usually writes and the content of each document represented by these topics. For these parameters, Gibbs sampling algorithms [112] and approximate inference with variational methods are usually used to estimate them.

In addition, McCallum et al. [113] proposed three Bayesian hierarchical models for identifying roles in mail data sets, namely author recipient topic model (ART), RART1 (role author recipient topic model), and RART2. ART model is a directed graph model of a word, which is derived from messages generated between a given author and a group of recipients. Because the roles to be identified exist in the subject groups (or lexical categories) of each meta-ancestor (author, recipient) associated with them, the ART model roles are unknown. On the basis of ART model, the author improves it to get two other models. In both models, roles, as a thousand-year-old random variable, are explicitly modeled in Bayesian networks. Therefore, a role is a mixture of themes that characterize the concerns of two people (the author and the recipient). In addition, in conference mining [114], the roles identified by ConMin model are usually specific expert or expert roles. In addition, in order to obtain a global analysis of the topics in scientific papers, additional dimensions such as time and sources are often added to the analysis. For example, semantics and temporal information-based maven search (STMS) model [115] can simultaneously obtain the author's potential topics, venues (meetings or journals), and time information and identify professional roles.

In a word, the traditional block model method is more related to graph theory and may be more suitable for role recognition in social networks. However, the block model is mainly constructed through the relationship structure of the network, ignoring the information content of user interaction. In social networks, users' roles are more or less expressed in the content generated by social media networks, or explicitly or implicitly expressed in the content generated by social media networks. In addition, in view of the current situation of the development of combinatorial optimization technology, block model method is slow and inefficient in many specific applications. For probabilistic models, this model can be said to be the first step to use more of their own information. In the process of role recognition, topic-based probabilistic model uses the key information of text between nodes, but lacks image block model to recognize from a global perspective. Therefore, how to effectively combine these two patterns is a challenge in current non-deterministic role recognition methods.

According to the identification model of unclear roles, the main identification basis and whether time factor is taken into account, the models described in this section are summarized as in Table 8.3.

Table 8.3 Uncertain role recognition models, based on time factors

Model	Year	Structure	Content	Interpretation
Stochastic block models	1987	√		Stochastic block model-structural equivalence
Stochastic block models	1996	√		Stochastic block model-structural equivalence
Block models	1993	√		Block model-regular equivalence
Stochastic block models	2004	√		Stochastic block model
ATM	2004		√	Author-theme model
LPCM	2007	√		Potential location model
MMSB	2008	√		Mixed member random block model
DMMB	2009	√		Dynamic mixed member block model
ConMin	2009		√	Conference mining model
STMS	2009		√	Semantic and time information

8.3.2   Role Recognition

This section will describe and summarize the current research status of influencer role, opinion leader role, and expert role identification methods in social networks at home and abroad. Finally, the status quo of other role recognition methods besides these three important roles is further described and summarized.

(1)   Influencer Role

At present, because of the rapid development of online social networks, such as Facebook, e-mail, online community, microblog, blog, WeChat, and so on, people have a broader space for social activities. But this kind of online social network may affect human activities in our real society. For example, a post posted by a star space can make others (especially fans or fans) respond or act crazily in their lives.

It is very important and interesting to identify the nodes that can affect the behavior of their neighbors from all the nodes in the online community network. In research and practice, it is becoming more and more popular to identify the role of influencers in online social networks and their applications in business. In October 2012, Facebook's online active users exceeded 1 billion, with more than 140 billion friends connecting. The explosive growth of data has aroused widespread concern in society. People pay more attention to identifying the role of influencers. In social networks, how to effectively identify these people or nodes that can mobilize others and expand their influence has also been constantly analyzed and studied. In this section, we will discuss the main methods, models, and various measurement methods for identifying the role of influencers in previous social networks.

Generally speaking, in the related research of social network, the method of identifying the role of influencers is mainly realized by measuring their influence or ranking them. From a technical point of view, it mainly includes technologies and methods based on network structure and content discovery.

In the identification method based on network structure, Kayes et al. [116] used network centrality matrix to analyze and concluded that blogs are usually connected by "core–edge" network structure. In this conclusion, influential blogs will be linked to become the core, while less influential blogs will be at the edge of the structure. Wu et al. [117] found a follower distribution approximating power law distribution and a friend distribution not obeying power law distribution in the microblog data studied and proposed XinRank algorithm to identify the role of influencers. Wen et al. [118] explained the existence of "peer" through homogeneity and extended PageRank algorithm to identify influential Twitter users and TwitterRank algorithm. Zhang et al. [119] took the validity of relationship strength as the criterion for identifying the role of influencers and concluded that identifying the role of influencers through the number of strongly connected users was better than other connections. Gliwa et al. [120] proposed a more reasonable and effective method of influencer role recognition based on the relationship between comments in blog topics and the definition of blog level.

In some social networks, only relying on the network structure characteristics may not be able to get real images between users, and users' content can enable us to understand the dynamic characteristics of the network. Therefore, recognition based on user content mining technology is also known as a more popular recognition method. Tang et al. [121] proposed a framework for detecting and identifying influential users from online healthcare forums. The framework first builds a social network in the healthcare forum. Then, an identification mechanism is established by combining link analysis, content analysis, and UserRank algorithm to identify influential users. Li et al. [122] proposed a marketing influential value (MIV) model to identify marketing value in blog data. The model is also based on two factors, network and content, to measure the dimension of blog characteristics and build an adaptive artificial neural network to identify potential influential blogs that support marketing or advertisers. Similarly, in blogs, Aziz et al. [123] use influence to identify influential blogs and propose an effective algorithm to identify influential blogs. These factors are mainly based on the semantic content of blog posts, quantitative analysis of the content of the posts, and comments in the posts by the followers.

Among the methods of identifying the role of influencers, researchers also put forward a variety of ways and indicators to measure the size of influence. For example, in measuring the impact of users, Moon et al. [124] proposed a weighted measurement based on homogeneity and vulnerability of blogs, established a quantitative impact model (QIM) to measure the impact score of blogs, and identified the role of influencers in blogs according to the impact score. Bui et al. [125] proposed a method to measure the impact of blogs in a community by using H-index, and to strengthen the calculation of information index, and finally identify the influential blogs. The H-index, proposed by Jorge Hirsch in 2005, refers to a person who has at most h papers cited h times. After the papers are listed in descending order of citation frequency, the H paper cited is shown in Fig. 8.3.

H-index is an index to measure academic achievement. It mainly measures the academic achievement and influence of a scholar or researcher by citing relationship. The higher a person's H-index, the higher his academic achievements, the greater his influence. Akritidis et al. [126] took the time factor of blog activities into account and proposed two ranking methods of blog influence. The two methods calculated the

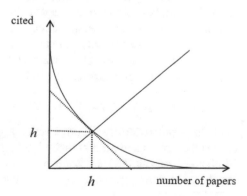

Fig. 8.3 H-index schematic diagram

score of blog posts based on specific indicators (such as the number of comments, the date of publication, and the age of posts). Identify influential blogs based on rankings. Moh et al. [111] built an improved model based on uniqueness and Facebook count by studying the existing models for determining influential blogs. The improved model can also measure the real influence of blogs by the number of links and comments, the time of posting and comments, the impact of reviewers and other factors, and rank bloggers accordingly, effectively identify influential blogs. Khan et al. [127] proposed a new metric for identification of influential bloggers (MIIB) to effectively identify the most influential blogs based on the production and popularity of blogs.

In addition, Ding et al. measured the influence of users through random walk of multi-relational data on microblog and proposed a method of combining random walk and multi-relational impact network to effectively identify the role of influencers in microblog. Akritidis et al. [128] identified blog roles with high productivity and high impact by introducing blog productivity index (BP-index) and impact index (BI-index). Shalaby et al. [129] identifies the role of influencers in Twitter users by identifying the factors that play a key role in the measurement of Twitter users' influence, verifying how different factors play a role in ranking influencers and how these factors are expressed as mathematical models. Also, Cai et al. [130] proposed a model to effectively mine the first k influential blogs, and Liu et al. [131] generated graph model using heterogeneous connection information and text content associated with nodes in the network. Among them, the direct influence of the related topics is mined by generating graph models, and then the indirect influence among network nodes is obtained. Although the author does not directly identify the role of the influencer, it provides us with a reference method to identify the role of the influencer through two kinds of influence. Similarly, Agarwal et al. [132] on the basis of studying what influential blogs are, from a blog Web site, assessed the various metrics collected to determine the impact of blog posts and proposed a preliminary model to quantify influential blogs, paving the way for finding various influential roles.

In short, the role recognition of influencers is more based on the social network topology and user content to identify and measure the impact of the size. In social networks, in the process of using graph theory to analyze network structure topology, it sometimes becomes more complex, especially today when the scale of social networks is getting larger and larger. In order to better identify the roles of influencers in online social networks, a new method including topological metrics, content weights, and link polarity values should be proposed in the future. In addition, the method based on user activity diffusion history is also effective in the identification process, and the extension of the method based on time and location should be an important aspect of future research [133].

(2) **Opinion Leader Role**

With the popularization of Internet technology, more and more people begin to like to get information and express their views on the Internet. Due to the uneven quality of Internet users and the concealment of the Internet itself, some users may express irresponsible views, opinions, or even statements that violate the law without being punished. Some of them may have a great impact on people's consciousness, and even some of them may endanger social order. Therefore, it is particularly important to properly guide the development of network public opinion. In the social network, opinion leaders play an important role in the generation and development of social network public opinion. Therefore, it is of great practical significance to identify and analyze the role of opinion leaders in social networks. In addition, the role of opinion leader is also an important research content in the field of network public opinion analysis. The role of opinion leader usually conveys personal opinions and attitudes to other network users and then influences and changes their opinions, attitudes, and decisions. The role of opinion leader referred to in this section refers to "activists" who provide information to others (e.g., publishing microblogs, posts, and replies, commenting on microblogs and posts published by other users) in social networks, exert influence on others, and act as intermediaries to accelerate information diffusion.

In recent years, in order to identify the role of opinion leaders in the network, researchers have conducted extensive research [134] and proposed many methods to identify the role of opinion leaders. Generally speaking, these methods are mainly based on user characteristics, content, and user interaction network for identification.

According to user characteristics and content recognition methods, Duan et al. [135] proposed a new method to identify opinion leaders from online stock message boards. Firstly, the method calculates the characteristics of user activities based on published information and then processes user data by clustering algorithm to generate clusters containing potential opinion leaders. Finally, through the method of emotional analysis, we analyze the relationship between users and emotions and the trend of real price changes and then identify the role of opinion leader. Hudli [136] constructs an online profile for each user on the basis of analyzing the characteristics of users' online activities and then uses K-means clustering algorithm to analyze these files or opinions, so as to identify opinion leaders in the forum. Yang et al. [137] proposed an opinion leader role recognition method based on text mining. This method identifies the role of opinion leader in the network by evaluating the professional characteristics of text content and the novelty and richness of information. Wang et al. constructed a multi-dimensional identification model based on the key factors of identifying the role of opinion leader, such as the number of users concerned, whether to verify identity, and the number of microblogs. They also proposed a formula for evaluating the importance of users. Finally, the model identifies the role of opinion leader in Weibo. Zhang et al. further analyzed the content-based "influence diffusion model" and then proposed a corpus-based ladder evaluation algorithm that can effectively identify the role of opinion leaders in online forums. In addition, Liu and his colleagues analyzed the characteristics of opinion leaders from the perspectives of the influence of network users and user activity, respectively, constructed an index system of opinion leaders by using analytic hierarchy process (AHP) and rough set decision analysis theory, and studied the cross-thematic nature of opinion leaders. Finally, they found that opinion leaders depend on themes.

Cheng et al. [138] proposed an IS_Rank algorithm for identifying opinion leaders in BBS based on user content and interactive network. The algorithm regards the user impact value (including content impact value and emotional impact value) as the weight of the link between users, which effectively improves the accuracy of opinion leader role recognition. Song et al. [139] proposed an algorithm based on link analysis and content analysis to solve the problem of ignoring the content of blog posts in traditional link analysis methods. The algorithm identifies the role of opinion leaders in blogs by calculating the impact score of bloggers (based on the number of internal links, the number of external links, the number of comments, and the length of articles). Similarly, Zhou et al. [140] proposed a method to identify the role of users in online social networks based on text mining and social network analysis. Luo et al. [134] proposed a hybrid data mining method based on interactive network and user characteristics to identify the role of opinion leader in microblog. According to Markov logical network, Zhang et al. [141] proposed a method of network opinion leader role recognition based on relational data. In addition, Fan et al. [142] proposed a new influence diffusion probability model (IDPM model), on which an open and inclusive network opinion leader screening model was established. Wang used intermediary centrality analysis method to identify opinion leaders in a quantitative way and constructed a microblog space opinion leader identification model. Yin and others obtain candidate opinion leaders by determining user centrality based on user relationship. Among them, the user-centered theory is based on the small-world network theory. Song et al. [143] proposed an InfluenceRank algorithm based on the importance of the node in the network and the novelty of blog information measured by cosine distance to identify the role of opinion leader in blog.

In fact, researchers at home and abroad pay great attention to PageRank in Leader-Rank algorithm in the research of opinion leader role mining. For example, Song et al. [144] used the familiar PageRank algorithm to rank users and then combined with the analysis of emotional orientation between users, the implicit relationship analysis of comments and the time attenuation of comments, finally excavated opinion leaders in complex networks. On the basis of introducing what opinion network is, Zhou et al. [145] proposed an opinion rank algorithm to identify the role of opinion leader in opinion network by improving PageRank algorithm to merge opinion scores. Ning et al. proposed an opinion leader recognition algorithm based on comment–response relationship by constructing a directed weighted comment network and drawing on the idea of PageRank algorithm.

PageRank algorithm calculates the ranking of Web pages according to the hyperlinks between Web pages. We generally regard PageRank algorithm as an algorithm for ranking the importance of Web pages. In fact, the algorithm draws lessons from traditional citation analysis. The basic idea of PageRank algorithm is that if there is a hyperlink between Web pages a and b and a points to b, then Web pages b will get the contribution value of Web pages a to it. The value depends on the importance of Web pages a, and the greater the importance, the higher the contribution value. Because links in Web pages are usually pointed to each other, the process of calculating contribution value is an iterative process. When the iteration reaches a certain stage, it will search and sort according to the score of Web pages, so as to get the final ranking results. The PR value (contribution value) of a Web page can be calculated by the following formula,

$$PR(a) = (1 - d) + d \sum_{i=1}^{n} \frac{PR(T_i)}{C(T_i)}$$

It is usually assumed that the initial contribution value of each Web page is 1, where $PR(a)$ denotes the contribution value of Web page a, and d denotes the random probability (usually 0.85) that a user visits a Web page. T_i is the other Web page pointing to page a, $C(T_i)$ is the number of Web pages, T_i pointing to the outer link, n is the general description of the Web page, and $\frac{PR(T_i)}{C(T_i)}$ is the contribution value of the linked Web page T_i of the Web page a. PageRank algorithm is a mature and common search engine page ranking algorithm. It is also often used in many disciplines such as literature statistics, social and information network analysis, and link prediction and recommendation.

In fact, there is a disadvantage when using PageRank algorithm. PageRank algorithm is not suitable for mining opinion leaders in complex networks with rapidly changing structures. In order to solve the shortcomings of PageRank algorithm in this respect, Lv et al. [146] proposed a LeaderRank algorithm for opinion leader mining. In the process of user sorting, the LeaderRank algorithm allocates a unit of resources for each node (except the base point, which is an additional point connected to each user through two-way links), and then equally allocates the resources to the neighbor nodes of the node until the equilibrium state is reached. This process is equivalent to free walking in direct network, which can be expressed by random matrix P, which is derived from formula $p_{ij} = a_{ij}/k_i^{out}$ and represents the probability of node i to node j in next free walking, where $a_{ij} = 1$ (when node i points to node j) or $a_{ij} = 0$ (when node i does not point to node j), k_i^{out} is the exit of node i. At time t, the score of node i is $s_i(t)$, while

$$s_i(t + 1) = \sum_{j=1}^{N+1} \frac{a_{ji}}{k_j^{out}}(t),$$

where the initial value of all nodes is $s_i(0) = 1$ (except the base point), and the initial value of base point is $s_g(0) = 0$.

Define the relationship value S as the user's final value,

$$S_i = s_i(t_c) + \frac{s_g(t_c)}{N},$$

where $s_g(t_c)$ is the value when the base point state is stable. Ultimately, according to the ranking of each user, the first user will be the opinion leader. Compared with PageRank algorithm, LeaderRank algorithm is more accurate in opinion leader mining and more stable in the presence of noise and malicious attacks. Xu et al. [147]

improved the LeaderRank algorithm proposed by Lv et al. by adding the emotional inclination and user activity, and the accuracy and anti-jamming ability of Leader-Rank algorithm were further improved. Xiao et al. [148] proposed a LeaderRank algorithm to identify opinion leaders in BBS based on community discovery (essentially a text categorization process) and emotion mining (building an emotion dictionary). Similarly, based on text analysis and emotion mining, Xiao et al. [149] combined the advantages of clustering algorithm and classification algorithm, proposed a method based on topic content analysis and interest group discovery, and then calculated the weight of user-to-user links by analyzing users' emotional tendencies to reply. On this basis, a new LeaderRank opinion leader discovery algorithm is proposed.

Opinion leader role is a common social network role in online social network, and it is also an important research topic in the field of social network research. At present, the role recognition mainly focuses on network analysis, text analysis, emotional analysis, and other methods. Most of these methods are based on a single node as the main research object, which can be analyzed and studied from the perspective of groups or groups in the future. In addition, because in social media, new opinion leaders will emerge at any time as the topic goes on or changes. Therefore, the opinion leaders can also be further analyzed and studied from the perspective of time and space.

(3) **Experts Role**

In recent years, in order to make full use of professional knowledge, expert discovery or identification has attracted considerable attention in the field of scientific research. In real life, expert discovery can also help us solve many challenging practical problems. For example, in order to improve the quality of published papers, peer experts are strongly recommended to review papers. For a given topic, how to judge the expertise of a given reviewer and how to match appropriately qualified reviewers are still a challenging problem.

In the field of social network expert role recognition, researchers have done a lot of research work, most of which use various measures to identify expert roles. According to its main recognition basis, it can also be divided into content-based identification method, link structure-based identification method, and the combination of the two methods. Among them, the concept of expert is generally defined as: in the social network, with the relevant knowledge of the topic under discussion, his opinions and ideas are credible, so I would like to make such a person an expert. The expert role recognition described in this paper mainly refers to the identification of expert roles in social networks, especially in network communities with professional knowledge.

Among the methods of social network expert role recognition, content-based method is the most important and common one. For example, Pal et al. found that experts preferred to answer such questions in their research on question and answer community (CQA) services [150]. We call this preferential choice problem selectivity bias and propose a mathematical model to estimate it. The research results show that the expert roles can be effectively identified from ordinary users by using the Gauss classification model and the selection bias of ordinary users. On this basis, Pal et al.

[151] continue to do further research in question answer community service (CQA). According to the questions that users choose to answer, they get the preference of users' preference and propose a probabilistic model based on preference. Finally, expert roles and potential expert roles are identified by machine learning. There are also research methods based on personal information as the center and document as the center to identify the relevant expert roles. In the method of personal information-centric identification, all documents and texts related to a potential expert are merged into a single personal file. Then, according to the given query corresponding to the file, the ranking score of each potential expert is estimated. Document-centric approaches analyze the content of each document separately, rather than creating a separate professional document. Some studies combine these two methods to improve the performance of expert role recognition effectively [152].

In addition, there is also a topic modeling method in the content-based expert role recognition method. Early Hofmann proposed a statistical implied semantics indexing (PLSI) algorithm [153], which was used to calculate the frequency of generating a word from a document based on an implied topic. However, the overfitting of PLSA model parameters is serious, and PLSI model cannot use a direct method to infer documents. Blei et al. [61] solved these problems by proposing a three-level hierarchical Bayesian model called Dirichlet distribution (LDA). On this basis, the authors' topic model and Author Conference Topic (ACT) model are proposed [154]. Each author is related to the polynomial distribution of a topic, words written or publishing conference. Theme models can calculate the correlation between potential experts and infer related topics, so that they can be ranked. In addition, Jung et al. [155] proposed a method to identify subject-centered expert roles from the metadata and complete text documents. This method extracts topic information from full-text content analysis to construct local classification, infers it to individuals and institutions in turn, and identifies subject-centered expert roles.

Link structure is another major aspect of expert role recognition. Adamic et al. [156] studied the patterns and behaviors of people asking and answering questions in Yahoo Answers Q&A forums and concluded that a user's link structure can better represent the topics of interest to the user than a user's profession. In specific identification methods, algorithms PageRank and HITS can be used to analyze the relationship between academic networks to find the role of authoritative experts. Some studies have proposed the expansion of PageRank and HITS algorithms. By measuring the weight of the cooperative relationship and the authority of users, the role of experts in the research community or problem-solving portal of digital libraries is identified, respectively [157]. Other studies have also obtained expert roles in bibliographic networks and co-citation networks by improving PageRank's algorithm [158]. In fact, these methods of identifying expert roles are all improvements in PageRank and HITS algorithm or other link-based methods to solve the limitations of some indicators (such as the number of citations), so as to get the ranking of bibliometrics, but the results of expert role recognition are often not the best.

The last major method of expert role recognition is to combine content-based and link-based structures [159]. However, this method cannot model all possible information in a unified way at the same time, but through local optimization of ranking in a limited subset of potential experts. For example, Campbell et al. [160] established a corresponding "professional map" by collecting all the mail related to a subject and analyzing the mail between each sender and recipient. Finally, all sender and recipient levels on this topic are obtained by improving HITS algorithm. Zhang et al. [161] first estimated the initial expert score of each potential expert through personal information (such as personal data, contact information, and publishing information) of the potential expert and selected the top potential expert to construct a subgraph. Then, a propagation-based method is proposed to improve the accuracy of expert discovery in subgraphs. In the academic network, Lin et al. [159] proposed a local weighted factor graph model (TWFG) by studying the problems found by subject experts in the citation network. The model combines the personal information and academic network information of potential experts in a unified way and improves the effectiveness of topic-level expert discovery.

In short, most of the current methods of expert identification often focus on personal information (such as topic-related and citation number) and network information (such as citation relationship). These methods often lead to some potential experts being ignored. Content-based expert role recognition mostly calculates the correlation between the query topics or inference topics of experts and users, but ignores the social relationship between experts. Other link analysis algorithms, such as PageRank and HITS, have a common topic drift problem in the process of identifying expert roles, and the expert roles they recognize are not the best. Therefore, based on the limitations of these two methods, we consider not only the relevance of a potential expert's special topic, but also the identification method of analyzing the network structure of potential experts, which is comparatively ideal for expert role recognition.

(4) **Other Roles**

In addition to the three most common social network roles mentioned above, many social network roles can be identified. For example, in online forums, Forestier et al. [162] proposed to get three kinds of social groups (subgroups) by principal component analysis (PCA). Then, maximize a subject standard and identify the celebrity roles (people who are most consistent with the subject) who participate in a subject from the subgroups of relevant users. In order to automatically identify celebrity roles in online forums, Forestier et al. [163] put forward some meta-criteria. A new data framework is constructed through three meta-standards and a baseline. Finally, it is concluded that the proposed meta-standard can successfully identify celebrity roles. In the social network, Xu et al. [164] construct an influence network by analyzing the information of users' social relations, ranks, and interactions in order to help enterprises to conduct marketing and enterprise management economically and effectively (such as reputation and publicity). This model not only maximizes the profit of enterprises, but also effectively identifies the role of the best value user (customer). Some studies separated communities under given conditions [165, 166], identified key person/player roles effectively from blogs, and given networks through clique percolation method (CPM) and fast modularity optimization (FMO) as well as interest-based multi-objective optimization. Some studies [167, 168] identified the WeiboRank method based on the quality and activity of Weibo and the machine learning methods through Bayesian classifier and backpropagation neural network, respectively, to identify high-value users and important user roles in Sina Weibo. In some studies [169], starters and followers are defined. Firstly, a method of random walk access disk sampling based on competition is proposed. Then, according to the measurement of network characteristics, the role of initiator and follower in social media is identified. Stadtfeld analysis [170] identifies potential behavioral roles in dynamic social networks by selecting behavioral patterns and clustering them according to their similarities. Lappas [171] and others proposed to use the optimal dynamic programming algorithm to identify the optimal effect nodes. The effect node here refers to the selection of a set of K active nodes under a given information propagation model, which can well explain the observed state of activity. Habiba et al. [172] obtained the conclusion that simple local measure (such as node degree) is the best index to judge the role of propagation blocking in the network by measuring some network topological structure characteristics and identified the best propagation blockers (spread blockers) in the dynamic network and the static network, respectively. Pal et al. [173] first depicted the characteristics of a group of social media writers in order to identify the most authoritative author roles under a given topic and then identified the authoritative author roles under a given topic through clustering and sorting.

In fact, there are many other roles in social networks besides those mentioned above, such as backbone nodes [174], positive value nodes [175], provocative role [176], network false information (rumor) expert role [177], and navy role [178]. Although these roles are not the main content in the current social network role recognition research, they are of great significance to our research and analysis of social networks and practical applications.

According to the types of other roles and the main identification methods and basis, the literature of other roles described in this book is summarized as shown in Table 8.4.

Table 8.4 Types of other roles and main identification methods, based on summary

Role name	Method/model	Structure	Content	Interpretation
Keyman	FMA, CPM, multi-objective	✓	✓	Faction filtering, fast modular optimization, multi-objective optimization algorithm of interest
High-value/important users	WeiboRank, machine learning-based	✓	✓	WeiboRank, machine learning method
Leaders and followers	Frequent pattern discovery, APPM, LUCI		✓	Three recognition methods
Initiators and followers	Random sampling	✓	✓	Random sampling
Potential behavioral roles	Individual patterns, clustering		✓	Individual patterns, clustering
Transmission interruption	Topology structure	✓		Topological structure
Most authoritative author	Clustering, sorting		✓	Clustering, sorting
Backbone node	FRT	✓		Forwarding relational tree algorithms
Positive value node	QTDG	✓		Quantitative time digraph
Agitator	Term frequency, cosine-correlation		✓	Two analytical methods
Online water army	WGM		✓	Hidden variable probability map model

References

1. Bailey NT (1978) The mathematical theory of infectious diseases and its applications. Immunology 34(5):955
2. Anderson R, May R (1992) Infectious diseases of humans: dynamics and control. Oxford University Press
3. Diekmann O, Heesterbeek JA (2000) Mathematical epidemiology of infectious diseases: model building, analysis and interpretation. Wiley
4. Boguñá M, Pastor-Satorras R (2002) Epidemic spreading in correlated complex networks. Phys Rev E Stat Nonlinear Soft Matter Phys 66(2):047104
5. Moreno Y, Gómez JB, Pacheco AF (2003) Epidemic incidence in correlated complex networks. Phys Rev E Stat Nonlinear Soft Matter Phys 68(2):035103
6. Pastorsatorras R, Vespignani A (2001) Epidemic dynamics and endemic states in complex networks. Phys Rev E Stat Nonlinear Soft Matter Phys 63(2):066117
7. Volchenkov D, Volchenkova L, Blanchard P (2002) Epidemic spreading in a variety of scale free networks. Phys Rev E Stat Nonlinear Soft Matter Phys 66(4 Pt 2):046137
8. Liu Z, Lai Y, Ye N (2003) Propagation and immunization of infection on general networks with both homogeneous and heterogeneous components. Phys Rev E Stat Nonlinear Soft Matter Phys 67(1):031911

9. Liu J, Wu J, Yang ZR (2004) The spread of infectious disease on complex networks with household-structure. Phys A 341(1):273–280

10. Barthelemy M, Barrat A, Pastorsatorras R (2004) Velocity and hierarchical spread of epidemic outbreaks in complex networks. Phys Rev Lett 92(17):178701

11. Newman ME, Watts DJ (1999) Scaling and percolation in the small-world network model. Phys Rev E Stat Phys Plasmas Fluids 60(6B):7332–7342

12. Moukarzel CF (1999) Spreading and shortest paths in systems with sparse long-range connections. Phys Rev E Stat Phys Plasmas Fluids Relat Interdisc Top 60(6 Pt A):R6263

13. Yang XS (2001) Chaos in small-world networks. Phys Rev E Stat Nonlinear Soft Matter Phys 63(2):046206

14. Yang XS (2002) Fractals in small-world networks with time-delay. Chaos Solitons Fractals 13(2):215–219

15. Li C (2004) Local stability and Hopf bifurcation in small-world delayed networks. Chaos Solitons Fractals 20(2):353–361

16. Chen D, Gao H, Lv L, Zhou T (2012) Identifying influential nodes in large-scale directed networks: the role of clustering. PLoS ONE 8(10):e77455

17. Ramani A, Carstea AS, Willox R (2004) Oscillating epidemics: a discrete-time model. Phys A 333(1):278–292

18. Hayashi Y, Minoura M, Matsukubo J (2004) Oscillatory epidemic prevalence in growing scale-free networks. Phys Rev E Stat Nonlinear Soft Matter Phys 69(2):016112

19. Li J, Zhang C, Tan H, Li C (2019) Complex networks of characters in fictional novels. In: IEEE international conference on computer and information science. Beijing, China, pp 417–420

20. Klemm K, Serrano MÁ, Eguíluz VM, Miguel MS (2012) A measure of individual role in collective dynamics. Sci Rep 2:292

21. Liu J, Lin J, Guo Q, Zhou T (2016) Locating influential nodes via dynamics-sensitive centrality. Sci Rep 6:21380

22. Ide K, Zamami R, Namatame A (2013) Diffusion centrality in interconnected networks. Proc Comput Sci 24:227–238

23. Min L, Liu Z, Tang X, Chen M, Liu S (2015) Evaluating influential spreaders in complex networks by extension of degree. Acta Phys Sin 64(8):387–397

24. Fowler JH, Christakis NA (2008) Dynamic spread of happiness in a large social network: longitudinal analysis over 20 years in the Framingham heart study. BMJ 337:a2338

25. Centola D (2010) The spread of behavior in an online social network experiment. Science 329(5996):1194–1197

26. Ugander J, Backstrom L, Marlow C, Kleinberg J (2012) Structural diversity in social contagion. Proc Natl Acad Sci 109(16):5962–5966

27. Cui P, Tang M, Wu Z (2017) Message spreading in networks with stickiness and persistence: large clustering does not always facilitate large-scale diffusion. Sci Rep 4:6303

28. Mislove A, Marcon M, Gummadi KP, Druschel P, Bhattacharjee B (2007) Measurement and analysis of online social networks. In: Proceedings of the 7th ACM SIGCOMM conference on Internet measurement (IMC 2007). ACM Press, New York, pp 29–42

29. Zhang J, Tang J, Zhuang H, Leung C, Li J (2014) Role-aware conformity influence modeling and analysis in social networks. In: Proceedings of the 28th AAAI conference on artificial intelligence. AAAI Press, Menlo Park, pp 958–965

30. Newman MEJ (2005) A measure of betweenness centrality based on random walks. Soc Netw 27(1):39–54

31. Sabidussi G (1966) The centrality index of a graph. Psychometrika 31(4):581–603

32. Borgatti SP (2005) Centrality and network flow. Soc Netw 27(1):55–71

33. Katz L (1953) A new status index derived from sociometricanalysis. Psychometrika 18(1):39–43

34. Kitsak M, Gallos LK, Havlin S, Liljeros F, Muchnik L, Stanley HE, Makse HA (2010) Identification of influential spreaders in complex networks. Nat Phys 6(11):888–893

35. Liu J, Ren Z, Guo Q (2013) Ranking the spreading influence in complex networks. Phys A 392(18):4154–4159

36. Zeng A, Zhang C (2013) Ranking spreaders by decomposing complex networks. Phys Lett A 377(14):1031–1035
37. Borge-Holthoefer J, Moreno Y (2012) Absence of influential spreaders in rumor dynamics. Phys Rev E 85(2):026116
38. Borge-Holthoefer J, Meloni S, Gonçalves B, Moreno Y (2013) Emergence of influential spreaders in modified rumor models. J Stat Phys 151(1–2):383–393
39. Liu Y, Tang M, Zhou T, Do Y (2015) Core-Like groups result in invalidation of identifying super-spreader by k-shell decomposition. Sci Rep 5:9602
40. Lv L, Zhou T, Zhang Q, Stanley E (2016) The H-index of a network node and its relation to degree and coreness. Nat Commun 7:10168
41. Berkhin P (2005) A survey on PageRank computing. Int Math 2(1):73–120
42. Kleinberg JM (1999) Authoritative sources in a hyperlinked environment. J ACM 46(5):604–632
43. Lv L, Zhang Y, Yeung C, Zhou T (2011) Leaders in social networks, the delicious case. PLoS ONE 6(6):e21202
44. Li Q, Zhou T, Lv L, Chen D (2014) Identifying influential spreaders by weighted LeaderRank. Phys A 404:47–55
45. Granovetter MS (1973) The strength of weak ties. Am J Sociol 78(6):1360–1380
46. Krackhardt D (1992) The strength of strong ties: the importance of Philos in organizations. Harvard Business School Press, Boston, pp 216–239
47. Shi X, Zhu J, Cai R, Zhang L (2009) User grouping behavior in online forums. In: Proceedings of the 15th ACM SIGKDD international conference on knowledge discovery and data mining (KDD 2009). ACM Press, New York, pp 777–786
48. Zhao Z, Yu H, Zhu Z, Wang X (2014) Identifying influential spreaders based on network community structure. Chin J Comput 37(4):753–766
49. Hu Q, Yin Y, Ma P, Gao Y, Zhang Y, Xin X (2013) A new approach to identify influential spreaders in complex networks. Acta Phys Sin 62(14):9–19
50. Burt RS, Kilduff M, Tasselli S (2013) Social network analysis: foundations and frontiers on advantage. Annu Rev Psychol 64:527–547
51. Han Z, Wu Y, Tan X, Duan D, Yang W (2015) Ranking key nodes in complex networks by considering structural holes. Acta Phys Sin 64(5):429–437
52. Lou T, Tang J (2013) Mining structural hole spanners through information diffusion in social networks. In: Proceedings of the 22nd international conference on world wide web steering committee. ACM Press, New York, pp 825–836
53. Yang Y, Tang J, Leung C, Sun Y, Chen Q, Li J, Yang Q (2014) RAIN: social role-aware information diffusion. In: Proceedings of the 29th AAAI conference on artificial intelligence. AAAI Press, Menlo Park, pp 367–373
54. Zhao X, Guo S, Wang Y (2017) The node influence analysis in social networks based on structural holes and degree centrality. In: IEEE international conference on computational science and engineering. Guangzhou, pp 708–711
55. Duan S, Wu B, Wang B (2014) TTRank: user influence rank based on tendency transformation. J Comput Res Dev 51(10):2225–2238
56. Fan X, Zhao J, Fang B, Li Y (2013) Influence diffusion probability model and utilizing it to identify network opinion leader. Chin J Comput 2:360–367
57. Song X, Chi Y, Hino K, Tseng B (2007) Identifying opinion leaders in the blogosphere. In: Proceedings of the 16th ACM conference on information and knowledge management. ACM Press, New York, pp 971–974
58. Agarwal N, Liu H, Tang L, Philip S (2008) Identifying the influential bloggers in a community. In: Proceedings of the 2008 international conference on web search and data mining. ACM Press, New York, pp 207–218
59. Bakshy E, Hofman JM, Mason WA, Watts DJ (2011) Everyone's an influencer: quantifying influence on twitter. In: Proceedings of the 4th ACM international conference on web search and data mining. ACM Press, New York, pp 65–74

60. Peng H, Zhu J, Piao D, Yang R, Zhang Y (2011) Retweet modeling using conditional random fields. In: Proceedings of the 11th IEEE international conference on data mining workshops. IEEE. IEEE Computer Society, Washington, DC, pp 336–343
61. Blei DM, Ng AY, Jordan MI (2003) Latent Dirichlet allocation. J Mach Learn Res 3:993–1022
62. Dietz L, Bickel S, Scheffer T (2007) Unsupervised prediction of citation influences. In: Proceedings of the 24th international conference on machine learning. ACM Press, New York, pp 233–240
63. Tang J, Sun J, Wang C, Wang C, Yang Z (2009) Social influence analysis in large-scale networks. In: Elder J, Fogelman FS (eds) Proceedings of the 15th ACM SIGKDD international conference on knowledge discovery and data mining. ACM Press, New York, pp 807–816
64. Li D, Shuai X, Sun G, Tang J, Ding Y, Lou Z (2012) Mining topic-level opinion influence in microblog. In: Proceedings of the 21st ACM international conference on information and knowledge management. ACM Press, New York, pp 1562–1566
65. Guo W, Wu S, Wang L, Tan T (2018) Social-relational topic model for social networks. In: Proceedings of the 24th ACM international conference on information and knowledge management. ACM Press, New York, pp 1731–1734
66. Weng J, Lim E, Jiang J, He Q (2010) TwitterRank: finding topic-sensitive influential twitterers. In: Proceedings of the 3rd ACM international conference on web search and data mining. ACM Press, New York, pp 261–270
67. Cui P, Wang F, Liu S, Ou M, Yang S, Sun L (2011) Who should share what? item-level social influence prediction for users and posts ranking. In: Proceedings of the 34th international ACM SIGIR conference on research and development in information retrieval. ACM Press, New York, pp 185–194
68. Tsur O, Rappoport A (2012) What's in a hashtag? content based prediction of the spread of ideas in microblogging communities. In: Proceedings of the 5th ACM international conference on web search and data mining. ACM Press, New York, pp 643–652
69. Gerrish S, Blei DM (2010) A language-based approach to measuring scholarly impact. In: Proceedings of the 27th international conference on machine learning. ACM Press, New York, pp 375–382
70. Shaparenko B, Joachims T (2007) Information genealogy: uncovering the flow of ideas in non-hyperlinked document databases. In: Proceedings of the 13th ACM SIGKDD international conference on knowledge discovery and data mining. ACM Press, New York, pp 619–628
71. Romero DM, Meeder B, Kleinberg J (2011) Differences in the mechanics of information diffusion across topics: idioms, political hashtags, and complex contagion on twitter. In: Proceedings of the 20th international conference on world wide web. ACM Press, New York, pp 695–704
72. Yang J, Leskovec J (2010) Modeling information diffusion in implicit networks. In: Proceedings of the 10th IEEE international conference on data mining. IEEE Computer Society, Washington, DC, pp 599–608
73. Wang Q, Mao Z, Wang B, Guo L (2017) Knowledge graph embedding: a survey of approaches and applications. IEEE Trans Knowl Data Eng 29(12):2724–2743
74. Trusov M, Bodapati AV, Bucklin RE (2010) Determining influential users in internet social networks. J Mark Res 47(4):643–658
75. Goyal A, Bonchi F, Lakshmanan LVS (2010) Learning influence probabilities in social networks. In: Davsion BD, Suel T (eds) Proceedings of the 3rd ACM international conference on web search and data mining. ACM Press, New York, pp 241–250
76. Xiang R, Neville J, Rogati M (2010) Modeling relationship strength in online social networks. In: Proceedings of the 19th international conference on world wide web. ACM Press, New York, pp 981–990
77. Zaman TR, Herbrich R, Van Gael J, Stern D (2010) Predicting information spreading in twitter. Proc Workshop Comput Soc Sci Wisdom Crowds 104(45):17599–17601
78. Yang Z, Guo J, Cai K, Stern D (2010) Understanding retweeting behaviors in social networks. In: Proceedings of the 19th ACM international conference on information and knowledge management. ACM Press, New York, pp 1633–1636

79. Cha M, Haddadi H, Benevenuto F, Gummadi PK (2010) Measuring user influence in twitter: the million follower fallacy. In: Proceedings of the 4th international AAAI conference on weblogs and social media. AAAI Press, Menlo Park, pp 10–17
80. Mao J, Liu Y, Zhang H, Ma S (2014) Social influence analysis for microblog user based on user behavior. Chin J Comput 37(4):791–800
81. Zhang J, Tang J, Li J, Liu Y, Xing C (2015) Who influenced you? Predicting retweet via social influence locality. ACM Conf Trans Knowl Discov Data 9(3):25
82. Romero DM, Galuba W, Asur S, Huberman B (2011) Influence and passivity in social media. In: Proceedings of the machine learning and knowledge discovery in databases. Springer, Berlin, pp 18–33
83. Bao P, Shen H, Huang J, Cheng X (2013) Popularity prediction in microblogging network: a case study on SinaWeibo. In: Proceedings of the 22nd international conference on world wide web companion steering committee. ACM Press, New York
84. Haveliwala T, Kamvar S, Jeh G (2003) An analytical comparison of approaches to personalizing PageRank. Technical report, Stanford
85. Tang J, Sun J, Wang C, Yang Z (2009) Social influence analysis in large-scale networks. In: Elder J, Fogelman FS (eds) Proceedings of the 15th ACM SIGKDD international conference on knowledge discovery and data mining. ACM Press, New York, pp 807–816
86. Xiang B, Liu Q, Chen E, Xiong H, Zheng Y, Yang Y (2013) PageRank with priors: an influence propagation perspective. In: Proceedings of the 23rd international joint conference on artificial intelligence. AAAI Press, Menlo Park, pp 2740–2746
87. Ding Z, Zhou B, Jia Y, Zhang L (2013) Topical influence analysis based on the multi-relational network in microblogs. J Comput Res Dev 50(10):2155–2175
88. Wang C, Guan X, Qin T, Zhou Y (2015) Modeling on opinion leader's influence in microblog message propagation and its application. J Softw 26(6):1473–1485
89. Jin Y, Xie J, Guo W, Luo C, Wu D, Wang R (2019) LSTM-CRF neural network with gated self attention for Chinese NER. IEEE Access 7:136694–136703
90. Huang J, Li C, Wang WQ, Shen H, Cheng X (2013) Temporal scaling in information propagation. Sci Rep 4:5334–5334
91. Li X, Zhang H, Zhang R, Liu Y, Nie F (2019) Generalized uncorrelated regression with adaptive graph for unsupervised feature selection. IEEE Trans Neural Netw Learn Syst 30(5):1587–1595
92. Zhao J, Yu L, Li J (2013) Node influence calculation mechanism based on Bayesian and semiring algebraic model in social networks. Acta Phys Sin 62(13):130201
93. Pan W, Dong W, Cebrian M, Kim T (2012) Modeling dynamical influence in human interaction: using data to make better inferences about influence within social systems. Signal Process Mag 29(2):77–86
94. Qin D, Zhou X, Chen L, Huang G, Zhang Y (2020) Dynamic connection-based social group recommendation. IEEE Trans Knowl Data Eng 32(3):453–467
95. Rodriguez M, Leskovec J, Balduzzi D (2014) Uncovering the structure and temporal dynamics of information propagation. Netw Sci 2(1):26–65
96. Laurent A, Camelin N, Raymond C (2014) Boosting bonsai trees for efficient features combination: application to speaker role identification. In: Conference of the International Speech Communication Association, Singapore, pp 76–80
97. Borgatti SP, Everett MG (1992) Notions of position in social network analysis. Sociol Methodol 22(4):1–35
98. Forestier M, Stavrianou A, Velcin J (2012) Roles in social networks: methodologies and research issues. Web Intell Agent Syst 10(1):117–133
99. Vladimir B (2005) Generalized block modeling. Cambridge University Press
100. Borgatti SP, Everett MG (1993) Two algorithms for computing regular equivalence. Soc Netw 15(4):361–376
101. Breiger RL, Boorman SA, Arabie P (1976) An algorithm for clustering relational data with applications to social network analysis and comparison with multidimensional scaling. J Math Psychol 12(3):328–383

102. Fienberg S, Wasserman S (1981) Categorical data analysis of single sociometric relations. Sociol Methodol 49(12):156–192
103. Holland PW, Leinhardt S (1981) An exponential family of probability distributions for directed graphs. Publ Am Stat Assoc 76(373):33–50
104. Snijders TA, Nowicki K (1997) Estimation and prediction for stochastic block models for graphs with latent block structure. J Classif 14(1):75–100
105. Handcock MS, Raftery AE, Tantrum JM (2007) Model-based clustering for social networks. J Roy Stat Soc 170(2):301–354
106. Zhang X, Wang C, Su Y, Pan L, Zhang H (2017) A fast overlapping community detection algorithm based on weak cliques for large-scale networks. IEEE Trans Comput Soc Syst 4(4):218–230
107. Airoldi EM, Blei DM, Fienberg SE (2008) Mixed membership stochastic block models. J Mach Learn Res 9(5):1981
108. Fu W, Song L, Xing EP (2009) Dynamic mixed membership block model for evolving networks. Ann Appl Stat 4(2):535–566
109. Steyvers M, Griffiths T (2007) Probabilistic topic models. Handb Latent Semant Anal 427(7):424–440
110. Griffiths RM (2016) The author-topic model for authors and documents. IEEE Conf Web Inform Syst Appl, IEEE, pp 487–494
111. Moh TS, Shola SN (2013) New factors for identifying influential bloggers. In: IEEE international conference on big data. IEEE conference, vol 1, pp 18–27
112. Griffiths TL, Steyvers M (2004) Finding scientific topics. Proc Natl Acad Sci USA 101(1):5228–5235.
113. Mccallum A, Wang X, Corrada-Emmanuel A (2010) Topic and role discovery in social networks with experiments on Enron and academic email. J Artifi Intell Res 30(2):249–272
114. Daud A, Li J, Zhou L (2009) Conference mining via generalized topic modeling. In: Proceedings of the European conference on machine learning and knowledge discovery in databases. Springer, pp 244–259
115. Daud A, Li J, Zhou L (2009) A generalized topic modeling approach for maven search. In: International Asia-Pacific web conference and web-age information management, Springer, pp 138–149
116. Kayes I, Qian X, Skvoretz J (2012) How influential are you: detecting influential bloggers in a blogging community. In: International conference on social informatics. Springer, pp 29–42
117. Wu X, Wang J (2014) Microblog in China: identify influential users and automatically classify posts on Sina microblog. J Ambient Intell Hum Comput 5(1):51–63
118. Weng J, Lim E, Jiang J (2010) TwitterRank: finding topic-sensitive influential twitterers. WSDM, ACM, pp 261–270
119. Zhang Y, Li X, Wang T (2013) Identifying influencers in online social networks: the role of tie strength. Int J Intell Inf Technol 9(1):1–20
120. Gliwa B, Zygmunt A (2015) Finding influential bloggers. Comput Sci 5(2):127–131
121. Tang X, Yang C (2010) Identifying influential users in an online healthcare social network. In: IEEE international conference on intelligence and security informatics, IEEE, pp 43–48
122. Li Y, Lai C, Chen C (2009) Identifying influential bloggers using blogs semantics. International conference on Frontiers of Information Technology, Islamabad, Pakistan, 1–6, 2010
123. Aziz M, Rafi M (2010) Identifying influential bloggers using blogs semantics. In: International conference on frontiers of information technology. Islamabad, Pakistan, pp 1–6
124. Moon EM, Han S (2010) A qualitative method to find influencers using similarity-based approach in the blogosphere. Int J Soc Comput Cyber Phys Syst 1(1):56
125. Bui D, Nguyen T, Ha Q (2014) Measuring the influence of bloggers in their community based on the H-index family. Adv Intell Syst Comput 282:313–324
126. Akritidis L, Katsaros D, Bozanis P (2009) Identifying influential bloggers: time does matter. In: IEEE/WIC/ACM International Joint Conference on Web Intelligence and Intelligent Agent Technology, vol 1, pp 76–83

127. Khan HU, Daud A, Malik TA (2015) MIIB: ametric to identify top influential bloggers in a community. PLoS ONE 10(9):e0138359
128. Akritidis L, Katsaros D, Bozanis P (2011) Identifying the productive and influential bloggers in a community. IEEE Trans Syst Man Cybernet Part C 41(5):759–764
129. Shalaby M, Rafea A (2013) Identifying the topic-specific influential users and opinion leaders in twitter. Acta Press (793), pp 16–24
130. Yichuan C (2009) Mining influential bloggers: from general to domain specific. In: Knowledge-based and intelligent information and engineering systems, Springer, pp 447–454
131. Liu L, Tang J, Han J (2010) Mining topic-level influence in heterogeneous networks. In: ACM conference, ACM, pp 199–208
132. Agarwal B, Nitin L (2008) Identifying the influential bloggers in a community, ACM, pp 207–218
133. Rabade R, Mishra N, Sharma S (2014) Survey of influential user identification techniques in online social networks. Advances in Intelligent Systems & Computing, 235, pp 359–370
134. Luo J, Xu L (2015) Identification of microblog opinion leader based on user feature and interaction network. In: IEEE conference on web information system and application, IEEE Conference, pp 125–130
135. Duan J, Zeng J, Luo B (2014) Identification of opinion leaders based on user clustering and sentiment analysis. IEEE/WIC/ACM international joint conference on web intelligence, IEEE conference, pp 377–383
136. Hudli S, Hudli A, Hudli A (2013) Identifying online opinion leaders using K-means clustering. In: International conference on intelligent systems design and applications, IEEE conference, pp 416–419
137. Yang X, Huang C (2019) The overview of text mining technique. Sci Technol Inform 33:82
138. Cheng F, Yan C, Huang Y (2013) Algorithm of identifying opinion leaders in BBS. IEEE International conference on cloud computing and intelligent systems. IEEE conference, pp 1149–1152
139. Song Z, Dai H, Huang D (2012) An algorithm for identifying and researching on opinion leaders in the blogosphere. Microprocessors 2012(6): 37–40
140. Zhou X, Liang X, Zhao J (2018) Structure based user identification across social networks. IEEE Trans Knowl Data Eng 30(6):1178–1191
141. Zhang W, Li X, He H (2014) Identifying network public opinion leaders based on Markov logic networks. Sci World J 2014(5):268592
142. Fan X, Zhao J, Fang B (2013) Influence diffusion probability model and utilizing it to identify network opinion leader. Chin J Comput 36(2):360–367
143. Song X, Chi Y, Hino K (2017) Identifying opinion leaders in the blogosphere. In: Sixteenth ACM conference on information and knowledge management. ACM conference, pp 971–974
144. Song K, Wang D, Feng S (2011) Detecting opinion leader dynamically in Chinese news comments. In: International conference on web-age information management. Springer, pp 197–209
145. Hengmin Z, Zeng D, Changli Z (2009) Finding leaders from opinion networks, In: IEEE conference on Intelligence and security informatics, IEEE, pp 266–268
146. Li L, Zhang Y, Ho Y (2011) Leaders in social networks: the delicious case. PLoS ONE 6(6):e21202
147. Jun X, Zhu F, Liu S (2015) Identifying opinion leaders by improved algorithm based on LeaderRank. Comput Eng Appl 51(1):110–114
148. Xiao X, Xia X (2010) Understanding opinion leaders in bulletin board systems: structures and algorithms. In: IEEE conference on local computer networks. IEEE Computer Society, pp 1062–1067
149. Yu X (2012) Networking groups opinion leader identification algorithms based on sentiment analysis. Comput Sci 39(2):34–37
150. Pal A, Konstan JA (2010) Expert identification in community question answering: exploring question selection bias. In: ACM international conference on information and knowledge management. ACM, pp 1505–1508

151. Pal A, Harper FM, Konstan JA (2012) Exploring question selection bias to identify experts and potential experts in community question answering. ACM Trans. Inf. Syst. 30(2):10
152. Serdyukov P, Rode H, Hiemstra D (2008) Modeling multi-step relevance propagation for expert finding. In: ACM conference on information and knowledge management. ACM, pp 1133–1142
153. Hofmann, T (1999) Probabilistic latent semantic indexing. In: International ACM SIGIR conference on research and development in information retrieval. ACM, pp 50–57
154. Tang J, Jin R, Zhang J (2008) A topic modeling approach and its integration into the random walk framework for academic search. In: Eighth IEEE international conference on data mining. IEEE conference, pp 1055–1060
155. Jung H, Lee M, Kang I (2008) Finding topic-centric identified experts based on full text analysis. International ISWC+ASWC workshop on finding experts on the web with semantics. Busan, Korea, pp 56–63
156. Adamic LA, Zhang J, Bakshy E (2008) Knowledge sharing and yahoo answers: everyone knows something. In: International conference on world wide web. ACM conference, pp 665–674
157. Jurczyk P, Agichtein E (2007) Hits on question answer portals: exploration of link analysis for author ranking. In: International ACM SIGIR conference on research and development in information retrieval. ACM, pp 845–846
158. Ding Y, Yan E, Frazho A (2009) PageRank for ranking authors in co-citation networks. J Assoc Inf Sci Technol 60(11):2229–2243
159. Lin L, Xu Z, Ding Y (2013) Finding topic-level experts in scholarly networks. Scientometrics 97(3):797–819
160. Campbell S, Maglio P, Cozzi A (2003) Expertise identification using email communications. In: The twelfth international conference, ACM, pp 528–531
161. Zhang J, Tang J, Li J (2007) Expert finding in a social network. Advances in databases: concepts, systems and applications. In: Proceedings of the international conference on database systems for advanced applications (DASFAA), Bangkok, Thailand, 9–12 Apr 2007, pp 1066–1069
162. Forestier M, Velcin J, Zighed D (2012) Analyzing social roles using enriched social network on on-line sub-communities. In: International conference on digital society, IEEE, pp 17–22
163. Forestier M, Velcin J, Stavrianou A (2012) Extracting celebrities from online discussions. In: International conference on advances in social networks analysis and mining. IEEE Computer Society, pp 322–326
164. Xu K, Li J, Song Y (2012) Identifying valuable customers on social networking sites for profit maximization. Expert Syst Appl 39(17):13009–13018
165. Zygmunt A, Brodka P, Kazienko P (2011) Different approaches to groups and key person identification in blogosphere. In: IEEE international conference on advances in social networks analysis and mining, IEEE Computer Society, pp 593–598
166. Gunasekara RC, Mehrotra K, Mohan CK (2015) Multi-objective optimization to identify key players in large social networks. Soc Netw Anal Min 5(1):1–20
167. Zhang G, Bie R (2013) Discovering massive high-value users from Sina Weibo based on quality and activity. In: International conference on cyber-enabled distributed computing and knowledge discovery. IEEE Computer Society, pp 214–220
168. Liu J, Cao Z, Cui K (2013) Identifying important users in Sina microblog. In: Fourth international conference on multimedia information networking and security, IEEE Computer Society, pp 839–842
169. Mathioudakis M, Koudas N (2009) Efficient identification of starters and followers in social media. In: EDBT 2009, international conference on extending database technology. Saint Petersburg, Russia, 24–26 Mar 2009, pp 708–719
170. Stadtfeld C (2012) Discovering latent behavioral roles in dynamic social networks. In: IEEE international conference on social computing and 2012 ASE/IEEE international conference on privacy, security, risk and trust. IEEE Computer Society, pp 629–635

171. Lappas T, Terzi E, Gunopulos D (2010) Finding effectors in social networks. In: Proceedings of the 16th ACM SIGKDD international conference on knowledge discovery and data mining. ACM, pp 1059–1068
172. Habiba H, Yu Y, Berger-Wolf TY (2010) Finding spread blockers in dynamic networks. In: Advances in social network mining and analysis. Springer, Berlin, pp 55–76
173. Pal A, Counts S (2011) Identifying topical authorities in microblogs. In: ACM international conference on web search and data mining, ACM, pp 45–54
174. Sun W, Zheng D, Hu X (2015) Microblog-oriented backbone nodes identification in public opinion diffusion. In: IEEE international conference on audio, language and image processing, IEEE, pp 570–573
175. Hong Q, Hao L, Yuan L (2014) Identification of active valuable nodes in temporal online social network with attributes. Int J Inf Technol Decis Making 13(04):1450061
176. Nakajima S, Tatemura J, Hara Y (2006) Identifying agitators as important blogger based on analyzing blog threads. Lect Notes Comput Sci 3841:285–296
177. Liang C, Liu Z, Sun M (2012) Expert finding for microblog misinformation identification. COLING 703–712:2013
178. Han Z, Xu F, Duan D (2013) Probabilistic graphical model for identifying water army in microblogging system. J Comput Res Dev 50(Suppl):180–186

Part III
Applications

Social computing with artificial intelligence is a computational theory and method oriented to social sciences and a bridge between social problems and computational technology based on artificial intelligence. Broadly speaking, its application can include many industries and fields involved in social sciences, such as e-commerce, biomedicine, public opinion of social government and anti-terrorism. It is difficult for us to elaborate the application examples of all fields in social computing in this book. Therefore, the author hopes to show readers how to apply the data and models mentioned in Part I and Part II to solve practical problems through several typical application examples that have been studied or are relatively familiar with.

Therefore, Part III is called *Applications*, which includes Chaps. 9–12. It introduces three applications in security and anti-terrorism, business applications, unsupervised handwriting recognition, and social network behavior and psychology. Let the reader provide more specific research ideas for solving the practical problems of online crowd social computing by using machine learning models.

Chapter 9
Social Computing Application in Public Security and Emergency Management

This chapter will introduce the reader to an application case in the field of public safety and emergency management, which uses the image information of social network to assist the emergency management organization in emergency decision making of vehicle hijacking. In the last section of this chapter, readers will be briefed on other applications in the field of public safety and emergency management.

9.1 Introduction

In recent years, a series of unexpected public safety emergencies have occurred at home and abroad, which have had a significant impact on people's lives. For example, the hijacking of Philippine tourist buses in August 2010, the loss of Malaysian Airlines passenger planes in March 2014, the killing of pupils by gangsters in October 2014 in Jiangxi, and the frequent bus attacks or hijackings, etc. The frequent occurrence of various public security incidents has brought serious challenges to the rescue work of governments at all levels in emergencies. Especially in the preparations for major international conferences such as APEC and G20, security practice has become an important part of the exercise. For example, during the APEC security practice on November 3, 2014, the team carried out a detailed and thorough practical exercise, taking the hijacking of public vehicles as an example. It can fully reflect the government's concern and attention to major incidents of public traffic hijacking safety. Therefore, the emergency management of such emergencies has become a focus of academic research.

In emergencies, the common characteristics of the car hijacking-type public security emergency are: the trapped or hijacked people are in a relatively closed place, and the emergency processing center outside the place cannot understand the specific internal information. As a result, it is impossible to make the most accurate judgment on emergency response, so there is a "broken ring" in the chain of emergency rescue links to obtain information. As in the 2010 Philippine bus hijacking, the robbers

© Tsinghua University Press 2020
X. Liang, *Social Computing with Artificial Intelligence*,
https://doi.org/10.1007/978-981-15-7760-4_9

asked to close all the windows and curtains in the car. The robbers communicated with the outside world only through the door. However, the emergency decision-making center in the outside cannot understand the internal situation of the carriage. Blind judgment leads to the intensification of the contradiction of rescue and assault, and the loss of the rescue time when the driver escapes and sends out information before starting the attack. Eventually, several hostages were shot by robbers, killing eight people and seriously injuring two.

It can be said that in such public security emergencies, it has been a difficult problem to understand the information inside the closed environment. During the APEC security drill just carried out, the team mainly adopted the method of understanding the interior environment of the carriage by unmanned helicopter, or by water supply and other acts to observe the situation in the car. But there are also many risks in this kind of behavior. For example, armed robbers cannot allow unmanned aerial vehicles to approach hijacked vehicles. At the same time, when the emergency center dispatches special personnel to deliver water, robbers will be highly alert to their behavior, while other conventional ways of understanding the interior of the carriage are also very limited.

However, with the rapid popularization of Internet and mobile interconnection, the hijacked person of handheld mobile communication equipment can become the most direct and effective information source to transmit information in the car. Through the advanced computer and network technology, give full play to the role of the trapped in the car and fully combine the massive network resources, so that the communication of information inside and outside the car becomes possible. Therefore, it can greatly enhance the external grasp of the situation inside the carriage and make up for the "broken link" on the chain of "rescue link" which is caused by the failure to understand the internal situation of the disaster.

At present, in the field of emergency management research, the massive information on the Internet, especially on public social platforms, has not been fully utilized. Many losses in the network may be that the precious scene images sent by the parties risking their lives do not give full play to their precious value. That is to say, even if there is valuable information to deal with unconventional emergencies in the network, the existing emergency management system does not have a complete and reasonable mechanism to achieve its effective monitoring and early warning function.

Based on this background and the shortcomings of emergency information collection and transmission, this chapter introduces an effective image acquisition and emergency treatment solution for unconventional emergencies, which fully complements the information source of emergency management field decision making and plays a very important reference value for emergency decision making [1].

9.2   Problem Description and Extraction

9.2.1   Qualitative Analysis of Car Hijacking Scenario

This chapter takes the bus hijacking in the Philippines as an example. Under the pressure of the robbers, all the curtains of the bus were closed and the doors were controlled by the robbers. Therefore, the whole environment of the car is closed and cannot be observed by the outside world, that is, there is a closed space scene (CSS). The decision-makers of emergency treatment cannot observe the environment inside the bus through conventional channels. The robbers only negotiate with the outside world through an easy-to-control space channel, so the robbers are in a very advantageous position, while the emergency treatment center is in a passive state because of the scarcity of information. In such a crisis situation, acquiring information I within CSS scenarios can help emergency decision-making centers change from passive to active.

At this time, the hostages who were hijacked inside the carriage risked their lives to take pictures of the inside of the carriage and uploaded them to social platforms such as microblog or micromail as fast as possible, forming very valuable data. Generally speaking, compared with other forms of data sources, image has many advantages, such as fast speed, large amount of information, convenient transmission, and so on. It is the most helpful information way for the outside world to understand the internal environment of the carriage. Specifically, at least information that the image information set $I(i_1, i_2, \ldots, i_N)$ transmitted to the outside through the mobile phone can include:

i_1 the number of people hijacked in the car;
i_2 geographical location of the robbed person;
i_3 age characteristics of the robbed person;
i_4 number of robbers;
i_5 weapons held by robbers;
i_6 the main location of the robbers;
i_7 environmental characteristics of the cabin;

The main decision-making value of the above image information to the emergency decision-making center includes: (1) Understanding and mastering the number of hijacked people in the carriage, estimating the severity, impact scope, and emergency level of the hijacking incident. (2) Mastering the geographical location of the hijacked people can help the emergency center to understand the main geographical direction of rescue operations, while preventing mistaken injuries in violent rescue operations. (3) Understanding the age distribution of the hijacked, if there are more elderly and children, it is necessary to speed up rescue and prepare adequate medical measures. In negotiations, priority should also be given to the release of the elderly and children. (4) Understanding the number of robbers and the weapons held by the robbers will help to understand the strength of the enemy and ourselves, increase the bargaining chip, and help to determine the threat level of the robbers. (5) The determination of the main location of the robbers is helpful to the sniper's targeting and judgment of the robbers. (6) The information of environmental characteristics in the carriage can help the decision-making center to understand the potential physical obstacles and rescue obstacles in the carriage, so as to avoid unnecessary or wrong rescue measures.

Therefore, it is indispensable and the most important link in the whole process of emergency information acquisition to effectively replenish the "broken link" of the rescue chain in emergency decision making and effectively transmit valuable image information from the inside of the train to the emergency command center and the emergency decision-making center on the spot. However, according to the current research and practice results, there is no reliable channel and mechanism to achieve such a process of information transmission [2]. As mentioned above, in practice, the usual method is to approach the hijacked vehicle by UAV or negotiation, but the information obtained by this method is very limited, and the practice is relatively difficult. Therefore, making full use of IT technology, this paper proposes an effective solution to "break the loop" in the chain of information transmission and rescue, which is the main research and discussion content of this paper.

9.2.2 Quantitative Description of Information Value of Social Networks

Assuming that the emergency decision-making center receives an early warning notice of an emergency at time t_1, because the information obtained at this time is incomplete, it can only judge and predict the state and level of an emergency at time t_2 ($t_2 > t_1$) according to the alarmer and general knowledge and adopt corresponding emergency plan according to this judgment. When the scheme is implemented to the time of $t_1 + \Delta t$ ($t_1 < t_1 + \Delta t < t_2$), the emergency decision center obtains the amount of information as I and grasps more accurate field information. Therefore, it can timely adjust the judgment of t_1 time and the emergency measures taken, and make more accurate emergency plans [3].

In this process, let $S = \{s_1, s_2, …, s_m\}$ denote the level set of emergencies, where s_j and s_k denote the jth and kth levels of emergencies, respectively, when $j < k$, $s_j < s_k$ denotes that the level s_j of emergencies is lower than s_k. Assuming that each level of emergencies corresponds to an emergency plan, the set of emergency plans is expressed as $X = \{x_1, x_2, …, x_m\}$, where x_j is the emergency plan for the emergency level s_j, and the corresponding start-up cost vector of the emergency plan is expressed as $C = \{c_1, c_2, …, c_m\}$, where c_j is the start-up cost of scheme x_j, and when $j < k$, $c_j < c_k$, that is, the lower the level, the lower the start-up cost.

Let a_j^o denote the effect of x_j on the disposal of s_o in emergencies. The value of a_j^o is as follows:

When $s_j \geq s_o$, $a_j^o > 0$ indicates that the scheme x_j can completely control and deal with the s_o level of emergencies and exceed the expected effect. The bigger the difference between s_j and s_o is, the bigger the corresponding a_j^o is, which indicates that the better solution x_j can deal with the s_o level of emergencies.

When $s_j < s_o$, $a_j^o < 0$ indicates that the scheme x_j cannot achieve the expected effect when dealing with sudden event s_o, the bigger the difference between s_j and s_o, the smaller the corresponding a_j^o, which indicates that the scheme x_j has a worse effect on dealing with sudden event s_o [4].

9.3 Social Computing in Emergency Problem

9.3.1 Emergency Decision Support Architecture for Bus Hijacking

Assuming that the computer server has crawled a large amount of images resources from SNSs in the emergency management system, and further an SNS vast images pool comes into being, which can be indicated as,

$$P = \begin{bmatrix} x_{11} & \cdots & x_{1M} \\ \vdots & \ddots & \vdots \\ x_{N1} & \cdots & x_{NM} \end{bmatrix},$$

where P represents a set of SNS images pool, and the P vector set concludes N rows, meaning that this images pool concludes N images, and each row represents the basic information of the corresponding image. Simultaneously, P vector set concludes M columns, meaning this images pool includes M types of image information. For instance, the first column concludes the global positioning system (GPS) information of the image, the second column contains the post-time, the third column indicates the transmitting port information, and the fourth column includes the identification information, etc. In general, the SNS images pool concludes consistent information for each image $\vec{p}_i = (x_{i_1}, x_{i_2}, \ldots, x_{iM})$.

In practice, for any emergency scene $\omega \in \overline{\Omega}$, we extract the essential information of it and filter each image $\vec{p}_i$ in the existing SNS images pool. Taking the Philippine tourist bus hijacking incident as an example ω, we should set the GPS information $x_{i1} =$ be near Reno grandstand in Manila Downtown, Philippine, $x_{i2} =$ be two hours earlier of the incident to instant. In this example, it is from 7:00 am, August 23, 2010, to present. In this way, we give the specific value to each factor of the scene ω image vector and sort out new images pool P' based on network auxiliary information. In a word, it extracts images with similar information like similar post-location, post-time, and so on.

Whereas the images pool P' still contains an amount of noises, it needs to further extract an images pool P'' with T images, which strong correlated to the Philippine tourist bus hijacking incident from pool P' with S images, and $S \leq T$, i.e., extract

$$\text{from } p' = \begin{bmatrix} x_{11} & \cdots & x_{1M} \\ \vdots & \ddots & \vdots \\ x_{S1} & \cdots & x_{SM} \end{bmatrix} \text{ and get } p'' = \begin{bmatrix} x_{11} & \cdots & x_{1M} \\ \vdots & \ddots & \vdots \\ x_{T1} & \cdots & x_{TM} \end{bmatrix}, S \leq T.$$

This process could further be converted into a problem of image recognition and classification, which means how to extract the images related to a specific incident only according to the image itself based on computer technology. For example, we should extract images related to the Philippine tourist bus hijacking incident among many images posted near Reno grandstand and from 7:00am, August 23, 2010, to present in this case. Therefore, it demands an effective monitoring and alert model to implement a specific selection from a lot of images.

After obtaining the effective images pool P'' with S images, the system will transmit pool P'' to the EDC on site, and the EDC further extracts valuable information $I(i_1, i_2, \ldots, i_N)$ from S images, which may contain i_1 (the number of abductees in the bus), i_2 (the position of abductees in the bus), i_3 (the gender and features of abductees), i_4 (the number of hijackers), i_5 (the weapons of hijackers), i_6 (the position of hijackers in the bus), and i_7 (internal environmental features of the bus), etc. The EDC can make the maximal use of information in emergency decisions.

In conclusion, the most significant and difficult step during the whole valuable information extraction process is that how to further distinguish the images relevant to a specific incident ω among plenty of preselected images. As a consequence, we propose an anomaly detection algorithm model in allusion to specific emergency incidents.

9.3.2   Anomaly Detection Algorithm Model Based on SIFT Features and SVM Classification

The key point of anomaly detection algorithm is to label the images automatically, and the fundamental difficult is that there is an obvious semantic gap between low-level visual feature extracted by computers and high-level semantic interpretation of image content by users. In order to eliminate the semantic gap, the existing approaches dedicate to build a mapping relation between visual features and semantic labels with the help of a set of artificial labeled training data and subsequently add correlated labels to new images according to the mapping relation. In particular, the widely acknowledged solution at present is to utilize scale-invariant feature transform (SIFT) descriptors to get image features and use SVM as the classifier.

In this paper, we make an innovative practice for the widely accepted algorithm to emergency anomaly detection and propose an anomaly detection algorithm model based on image data aimed at specific emergency scenes. The method and the technology route of each module will be introduced next.

(1) Image Features Extraction Based on SIFT

SIFT is a local feature descriptor in the field of image processing, which was proposed by David G. Lowe in 2004. The algorithm shows strong robustness in the scale of image scaling, rotation, transformation, even brightness changes and affine transformation. There are four steps in SIFT feature extraction algorithm.

1. Extremum detection in scale space

We subsample the images repeatedly so that it can get a series of images of a pyramid. The definition of two-dimensional Gaussian filtering function is

$$G(x, y, s) = \frac{1}{2\pi\sigma^2} \exp^{\frac{-(x^2+y^2)}{2\sigma^2}}$$

where σ represents the variance of Gaussian function.

An $N \times N$ image $I(x, y)$ can be expressed in different spatial scales, which is named Gaussian image. The Gaussian image is obtained by a convolution between an image and Gaussian kernel,

$$L(x, y, s) = G(x, y, s) * I(x, y),$$

where σ is called scale space factor. The greater the value of σ is, the smoother the image is. Large scale corresponds to an overview of the image, while small scale corresponds to details of the image. DoG operator is defined as,

$$D(x, y, s) = [G(x, y, k^s) - G(x, y, s)] * I(x, y).$$

In order to detect the local extremum points of $D(x, y, \sigma)$, it demands a comparison between each point in DoG scale space and 26 points adjacent to its scale and position one by one. If pixel (x, y) is a possible SIFT key point, it must be an extremum point among the ambient 26 neighboring pixels (9 points in the last scale + 8 points in the same scale + 9 points in the next scale). All the local extremum points constitute a SIFT key points of alternative set.

2. Key point orientation

There are another two steps for all key points obtained from extremum detection to confirm its validity: the first step is to inspect an obvious discrepancy among its surrounding pixels, and the second step is to get rid of unstable edge response points (for there is a strong edge response effect for DoG operator).

3. Matching key points' size and direction

For the sake of operators' rotation invariance feature, it determines the main direction with the help of gradient histogram. The module value and direction of gradient at point (x, y) can be calculated as

$$m(x, y) = \sqrt{(L(x + 1, y) - L(x - 1, y))^2 + (L(x, y + 1)) - L(x, y - 1))^2},$$

$$\theta(x, y) = \tan^{-1} \frac{L(x, y + 1) - L(x, y - 1)}{L(x + 1, y) - L(x - 1, y)}$$

It should take into account each key point's gradient direction in its neighboring window. However, the peak of the histogram represents the neighboring gradient's main direction of the key point, as well as the main direction of the key point itself. We assign a direction parameter for each key point so that the operator has the feature of rotation invariance.

4. Generate SIFT descriptor

In order to guarantee the rotation invariance feature, firstly, the coordinate axis should be rotated to the direction of the key point. We take an 8×8 window centered on a key point, segment the window into 2×2 child windows, and then collect the orientation histogram of each child window.

The direction of each child window is decided by its 4×4 small blocks with the method above. However, the direction of each key point in the image is determined by $2 \times 2 = 4$ seed points' directions, and there is 8 directions information for each seed point, therefore there are $4 \times 8 = 32$ dimensions for each key point.

Each image becomes a set of many SIFT key points after being extracted by SIFT algorithm. Furthermore, each key point equals to a 128-dimensional feature vector and depicts the feature information of a portion of the objects in the image, such as the borders of an object and image gray change border. However, the number of key points for each image generated by SIFT algorithm is diverse, while it demands an integrated and standardized characteristic form for image tag library applied to image processing in emergency management system. Therefore, we bring a K-means algorithm into use for SIFT features clustering and standardize the clustering with the Euclidean distance. The distance between SIFT feature vector X and the ith clustering center is as,

$$D_i = \sqrt{\sum_{j=1}^{128} (x_j - k_{ij})^2},$$

where x_j represents the jth dimension of vector X, and k_{ij} is the jth dimension of the ith clustering center. Multiple iterative calculations of K-means algorithm are used to get the K clustering center, in which the value of K is determined by an overall consideration of multiple tests as well as clustering speed and classification accuracy.

For an image with N SIFT key points, firstly we analyze the distribution of N key points on the K clustering centers so that could bring the corresponding distribution vector feature of this image into being. The specific jth feature distribution vector is calculated as,

$$v_j = \frac{\sum_i^N s_{ij}}{N},$$

where $s_{ij} = \begin{cases} 1, \text{the ith key point in the jth center} \\ 0, \text{else} \end{cases}$.

In conclusion, the algorithm converts the image with N SIFT key points into a K-dimensional feature distribution vector, which represents the classification feature of the image and further becomes the feature vector of SVM algorithm next.

(2)   Image classification based on SVM classifier

SVM is a dichotomy classifier algorithm preponderant in small sample set, nonlinear and high dimensions identification proposed by Vapnik in 1995. In the process of concrete realization, we divide the procedure into two portions of "offline images training" and "online images discrimination."

For the offline images training part, we collected plenty of images involving bus hijacking related images as well as irrelevant ones, we divided and labeled them manually, then preprocessed the images with SIFT and K-means, and finally trained the SVM classifier with them separately.

Then for the online images discrimination part, we mixed the related and irrelevant images together, after preprocessing of SIFT and K-means, we classified them with the already trained SVM classifier, and then made statistic and analysis on the results to evaluate the effect of our algorithm.

Furthermore, the model can be further extended in other scenes. It collects images, respectively, according to different scene ω in scenes set $\overline{\Omega}$, then trains the corresponding SVM classifier so that it can constitute a more comprehensive emergency anomaly detection model for abundant scenarios.

(3)   Optimized emergency management solution to bus hijacking

We took our proposed emergency decision support architecture into an accurate situation of a tourist bus hijacking incident to prove it more feasible (see Fig. 9.1) [1].As illustrated in Fig. 9.1, tourists inside the bus uploaded the interior circumstance images to Twitter or Facebook, etc. The real-time emergency management system collected variety of images timely and stored them to the database. When the emergency decision center needed the inner condition images, they logged in the system based on our emergency decision support architecture, set the specific factor according to the events' situation, and finally picked up desirable images after SIFT–K-means–SVM disposal. However, the essential images will be transmitted to emergency decision center and be helpful to the decision.

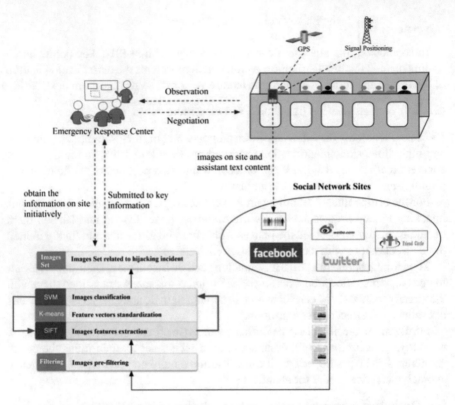

Fig. 9.1 Optimized emergency response process based on access to effective images information

References

1. Shen H, Liang X, Wang M (2016) Emergency decision support architectures for bus hijacking based on massive image anomaly detection in social networks. In: IEEE international conference on systems, man, and cybernetics, pp 864–869
2. Ceron A, Curini L, Iacus SM (2014) Every tweet counts? How sentiment analysis of social media can improve our knowledge of citizens' political preferences with an application to Italy and France. New Media Soc 16(2):340–358
3. Mayer-Schönberger V, Cukier K (2013) Big data: a revolution that will transform how we live, work, and thinking. Houghton Mifflin Harcourt, Boston, pp 50–72
4. Carneiro HA, Mylonakis E (2009) Google trends: a web-based tool for real-time surveillance of disease outbreaks. Clin Infect Dis 49(10):1557–1564

Chapter 10
Social Computing Application in Business Decision Support

This chapter will introduce readers to an application case of stock market investment decision-making based on the analysis of news content of online listed companies in the field of business decision-making support. In the last section of this chapter, we will give readers a brief introduction to other business decision support applications.

10.1 Introduction

With the rapid development of Internet and information technology, people have become more and more convenient and rapid to access information, and relevant information about listed companies is no exception. Among them, there are not only structured data such as stock price, turnover, and company financial situation but also unstructured data such as news, announcement, policy, especially unstructured data, which is particularly inflated. The structured data and unstructured data of listed companies are not only essentially different in form but also in the way of dissemination. The generation and acquisition of structured data of listed companies often have strong regularity, such as the price information and trading volume of listed companies' shares, which can be obtained at the end of each trading day, and financial data can be obtained at the disclosure date. Unstructured data are different, such as public announcements, news and policy changes of listed companies, often with strong unpredictability, a public announcement or news is often issued suddenly.

In addition, structured data and unstructured data have different degrees of difficulty in application. There are a large number of mature data models dealing with structured data of listed companies—CAPM calculation of expected return, data mining algorithm in artificial intelligence to predict stock prices, and so on. However, for unstructured data with a large amount of high-quality information, due to the lack of mature models, how to use them efficiently has always been a difficult problem. The unstructured information may sometimes be more valuable than structured data.

© Tsinghua University Press 2020
X. Liang, *Social Computing with Artificial Intelligence*,
https://doi.org/10.1007/978-981-15-7760-4_10

Unlike other unstructured data (e.g., hearsay, forum comments, microblog), the announcement of listed companies has two characteristics: authenticity and importance (legal constraints make it true, and legal requirements need to be disclosed because of its importance), which determine that the information it contains is of high quality. On the one hand, the accuracy of news and social media information is not guaranteed at all. On the other hand, there is a great possibility of repetition, redundancy, irrelevance, and unimportance, which results in a lot of noise.

Until the resumption of IPO in 2014, there were more than 2900 listed companies in mainland China. They traded on Shanghai Stock Exchange and Shenzhen Stock Exchange, respectively. Every day, hundreds of announcements were generated. It is not difficult to find that the amount of announcement information is very large, but not every piece of information has value. If we can find out valuable information from the massive information, and understand its impact on the stock price of the corresponding listed companies, or whether the issuance of an announcement has an impact on the stock price of the company on that day. From the perspective of investors, on the one hand, it is conducive to the discovery of investment opportunities. On the other hand, it also has a prediction of risk, which has a strong practical significance.

In the field of social computing, the core method of the existing research on financial decision support using unstructured data is to transform unstructured data into structured data first, and then use this structured data as an input of data mining method for decision support. Geva and ZahaviL [1] convert different types of text data into emotional values through emotional analysis, and then use neural networks, decision trees, and other algorithms to classify stocks into two categories: rising and non-rising. Based on this, a set of stock recommendation mechanism is proposed. Some studies [2, 3] used a representative text vector to represent the announcement, and used naive Bayesian, SVM, neural network, and k-nearest neighbor algorithms to predict the possible risk of listed companies' stock. The traditional financial field has a longer history of discussing issues related to announcement and stock price, which belongs to the category of event study. Beck et al. [4] concluded that the stock market would overreact to negative news by studying the fluctuation of certain time windows before and after news release. Muntermanns et al. [5] studied the impact of specific announcements on price and volume, and concluded that announcements have greater impact on smaller companies. Methodologically, in the processing and analysis of unstructured text data, word co-occurrence-based method has been widely adopted, and related text classification, support vector machine is also a more recognized method by scholars.

By synthesizing the related research in decision support field and traditional financial field, it can be seen that the decision science field mainly carries out text analysis first, transforms unstructured data into structured data, such as emotional value or uses space vector model to represent a document, and then acts as the input or part of the input of data mining algorithm. However, the traditional financial research often has relatively mature models to deal with structured data, and for unstructured data, there is a problem of insufficient utilization. It is true that traditional financial research tends to focus on some deep-seated issues, such as considering the different size of companies, different types of announcements will have different effects on the stock market.

We hope to consider not only the method of using unstructured data in the field of social computing but also the idea of considering the impact of news types on stock prices in the financial field. That is to say, we propose a method of dividing news types, and consider the difference of the impact of different types of news on stock prices, so as to achieve the purpose of decision support. This part refers to the research of Ma.

10.2 Social Computing in Business Decision Support

10.2.1 Model Overall Framework Diagram

In the general framework of the model of automatic reading and decision support system, when the network produces new news, the word segmentation system automatically partitions it. According to the keyword library established before, the news is expressed as a keyword vector. The classifier trained according to the historical news data before input is used to output the news category. Event research system gives empirical judgment of the news's impact on the stock price of listed companies based on the event research results obtained from historical transaction data. When new news enters the system, the keywords database, historical news database, and historical transaction database are updated, and the support vector machine classifier is trained again.

10.2.2 News Type Division Model

(1) Participle

In order to transform unstructured text data into structured data, word segmentation is a very effective means. This case chooses ICTCLA, a word segmentation system of Chinese Academy of Sciences. Word segmentation system judges that several words or words should be separated or formed into an inseparable meaningful word according to the dictionary, which has a good effect on general texts. However, the announcement news involved in this case belongs to a specific field. The original system incorrectly divides a meaningful word into several separate words, which affects the effect of word segmentation. For example, "Zengzhi," if no processing is done, the word segmentation system will divide the word into "Zengzhi" and "Zhizhi."

Another example is Guanghui Energy, which is the name of a listed company. Considering this influence and the good expansibility of ICTCLAS system, we introduce Sogou financial dictionary (including 12,239 words) and listed company name dictionary (containing 2912 words) into the system dictionary to enhance the system's ability to recognize specific professional words.

(2)   Keyword Extraction

One of the purposes of this case is to distinguish bulletin news according to different categories, and further, that is to say, to distinguish bulletin news according to different themes. The result of word segmentation is one of its bases. Analyzing the results of word segmentation, we find that if we want to define or divide a topic with the appearance or co-occurrence of some specific words, we should choose some representative words. For example, because the sample of this paper is the announcement news of listed companies in the afternoon, so almost every document will have such words as company, announcement, etc. They will appear not only in every document with a certain frequency but also in every document, so there is no distinction. In a specific topic, such as executive increase or decrease, there will be some words that appear steadily, such as "reduction," "circulation stock," "accumulation," and so on. These words will appear or even not appear in other topics at a very low frequency. These words, we call them keywords. The purpose of this part is to find out the key words by some means in order to classify the announcement news. tf-idf (word frequency-document frequency) can extract keywords by controlling word frequency and document frequency.

(3)   Vector Space Model Represents Text

Combined with the content of the space vector model, we use a text vector to represent each announcement news, that is, N keywords obtained above. In this way, every news item is an n-dimensional vector. For the value of each dimension, the general practice is to express the number of times the word represented by the dimension appears in the document.

Consider the use of k-means clustering in this paper, that is, the distance between documents (Euclidean distance is used in this paper). If the occurrence frequency is used to represent the value of each dimension, there may be a big deviation in calculating Euclidean distance. For example, there are 10 words in document 1 and 7 words in document 2, so there is a large distance difference in this dimension. In this paper, we are more concerned about whether there are some specific keywords or the co-occurrence of several keywords than the number of occurrences. Based on this, we define that the meaning of each dimension is whether there are keywords represented by that dimension in the document, and the value of that dimension is 1 when it appears, and vice versa, 0.

(4)   Clustering and Classifier Construction

After each news is represented by an n-dimensional 0–1 keyword vector, k-means algorithm is used to cluster the samples. The main purpose of clustering here is to get the initial types of news from the perspective of text mining, so as to lay the foundation for the next step of classifier construction. Therefore, when choosing the appropriate K value for clustering results, we should fully consider the rationality of the practical significance of news types.

After getting the main types of noon announcement news of listed companies, we label all the noon announcement news. So far, each news item is represented by a keyword vector with a category tag. Based on this, the SVM classifier is obtained by using the training samples of SVM.

In order to investigate the classification effect, three kinds of common kernel functions of support vector machine are experimented, respectively. At the same time, the samples are divided into training set and test set for experiment. Only when the accuracy and recall rate of the existing sample test set are high, can the news type of the new sample be predicted. At the same time, for multi-type support vector machine classification problem, this paper adopts the common method of dealing with multi-classification problem "one-to-one" to construct multiple classifiers, so as to get the classification results.

10.2.3   Event Research Model

(1)   Event Study and Abnormal Return Rate

The impact of the noon announcement on stock prices in this case study falls into the category of event study in the financial field. Event study method refers to the use of financial market data to determine the impact of a specific economic event on the value of a listed company. Here, specific economic events can be the release of favorable policies, the dissemination of corporate news, the disclosure of corporate matters, and so on.

The basic idea of the event study method is to set the event window as the time period during which the event affects, calculates the daily abnormal rate of return (the difference between the actual rate of return in the period and the rate of return in the event-free condition) and the cumulative abnormal rate of return, and measures the significance of the event impact with the statistical test of the two indicators.

Real yield is easy to calculate, as long as the price at a given time can be calculated. The key is the expected rate of return without events, which needs to be predicted or estimated. One simple method is to take the mean value in the window before the event occurs, and the other is to calculate the expected return using the CAPM model.

(2)   Label Determination

Unlike the typical event study, this case focuses on the market reaction on the day after the event, hoping to find out a set of response patterns, so it does not pay attention to whether there has been a response before.

When dealing with abnormal returns and saliency, this case regards abnormal returns as significant returns if the historical returns obey normal distribution and the returns exceed one standard deviation of the average returns in the past year.

After the announcement news was issued in the afternoon, the samples with abnormal returns in the afternoon were considered to be significantly affected by the announcement, and were labeled as 1. Otherwise, they were labeled as 0, that is to say, all samples were divided into two categories.

References

1. Geva T, Zahavi J (2014) Empirical evaluation of an automated intraday stock recommendation system incorporating both market data and textual news. Decis Support Syst 57(3):212–223
2. Wang L, Song R, Qu Z, Zhao H, Zhai C (2018) Study of China's publicity translations based on complex network theory. IEEE Access 6:35753–35763
3. Cui L, Bai L, Rossi L (2018) A preliminary survey of analyzing dynamic time-varying financial networks using graph kernels. In: Joint IAPR international workshops on statistical techniques in pattern recognition. Springer, Cham, pp 237–247
4. Klobner S, Becker M, Friedmann R (2012) Modeling and measuring intraday overreaction of stock prices. J Bank Finance 36(4):1152–1163
5. Muntermann J, Guettler A (2007) Intraday stock price effects of ad hoc disclosures: the German case. J Int Financ Mark Inst Money 17(1):1–24

Chapter 11
Social Computing Application in Unsupervised Oracle Handwriting Recognition Based on Pic2Vec Image Content Mapping

11.1 Image Feature Extraction Method

Image feature extraction is a basic concept in computer image processing. It refers to the extraction of image information by computer to determine whether each image point belongs to an image feature. The final result is to divide the points on the image into different subsets, which are often isolated points, continuous curves, or continuous regions. Image features can be extracted and represented from the edge, angle, region and ridge of the image. At present, the commonly used image features include color features, texture features, shape features, and spatial relationship features.

In this section, an image feature extraction method based on Pic2Vec is designed. First of all, through the gray processing method of transforming the color image containing brightness and color into the gray image, the image becomes an image containing only brightness information but not color information. Then the image content is processed by binary method, in which the pixel points with gray value greater than or equal to 120 are represented by 1, and the points with gray value less than 120 are represented by 0. Thus, the image content can be mapped to a simple representation of 0 and 1. Secondly, based on the simple representation of content mapping, the accumulation vectors of different directions can be obtained by superposition of multiple directions, which can be regarded as the simple feature extraction or representation of image content in this direction. Finally, we combine all direction vectors into a row vector, which is the feature representation of the image content.

The main process of image feature extraction method in this section is as follows:

Input Picture set Img $= \{\text{img}_1, \text{img}_2, \ldots, \text{img}_n\}$.
Process
1. Image graying processing, img.convert("L").
2. Binary image processing, img.point (lambda x: 1 if $x > 120$ else 0), map image content to 0 and 1 representations.
3. Extract horizontal and vertical feature vectors of image (hv, lv), Sumhlvector(img).

© Tsinghua University Press 2020
X. Liang, *Social Computing with Artificial Intelligence*,
https://doi.org/10.1007/978-981-15-7760-4_11

4. Extract image skew feature vector, SlantVector(img).
5. Combine picture feature vector, hb = np.concatenate(a, b, c).
Output representation of picture features, concateVec(mat).

For example, in a 32 × 32 (pixel) image, the image content is mapped to a simple representation of 0 and 1 through image gray processing.

As we mainly focus on the recognition of oracle bone inscriptions (Chinese), considering that Chinese characters are mainly composed of horizontal and vertical strokes, then we simply stack them horizontally, vertically, and obliquely,

1. Add it horizontally to get a 32 × 1 vector.
2. Equally, add it vertically to get a 32 × 1 vector.
3. Tilt a 45° angle and add it to the edge to get a 45° vector.
4. Tilt a 135° vector and get a 135° vector.

Finally, the four vectors are combined to get a feature representation vector of an image. We regarded this vector as a simple feature representation of the image content.

11.2 Experiments and Analysis

In order to verify the image feature representation method, we used the image visualization method to reduce the dimension through manifold learning and present the image feature vectors with different contents in the graphics. Among them, manifold learning is mainly to redraw image features from high-dimensional space into a low-dimensional space. We verified it by handwritten 0, 3, 5, and 9 pictures.

Through unsupervised learning, after clustering, manifold learning can reduce the visual effect of dimension, as shown in Fig. 11.1. In Fig. 11.1, different gray levels represent different clusters.

As can be seen from Fig. 11.1, through the feature representation method proposed in this section, the feature vectors of different content pictures have obvious feature differences, which can effectively distinguish different image contents.

At the same time, we conduct simple training and classification through machine learning model (logistic regression model), and the results are shown in Table 11.1.

As can be seen from Table 11.1, the image feature representation method proposed in this section can effectively achieve feature extraction or representation of different categories of image content. In the traditional classification model, it has a good classification effect.

Two decades ago, machine recognition of handwriting hovered between 80 and 90% of the time. However, after the emergence of deep learning CNN, amazing progress has been made, and the recognition rate leaped to over 99%. As long as there are enough training data, deep neural network has a strong ability to extract features under the condition of supervision, and it is not difficult to solve the problem of handwritten machine recognition. However, CNN basically a supervised learning,

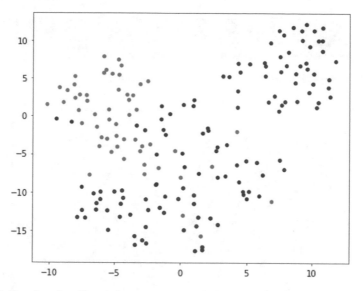

Fig. 11.1 Visual results of image feature manifold learning and dimensionality reduction

Table 11.1 Experimental results of image classification

Class	Precision	Recall	f1-score
0	0.96	0.89	0.93
3	0.89	0.89	0.89
5	0.83	0.88	0.86
9	0.89	0.94	0.91
Average	0.89	0.90	0.90

that is to say, some samples need to be labeled. Sometimes, it is not possible in practice. At present, using unsupervised machine learning to complete handwriting recognition is still a problem, and that is the recognition rate is not high. In this section, we carry out an attempt.

First, we write 100 Arabic numeral 5 with left hand and right hand, respectively. Some examples of left hand and right hand are shown in Figs. 11.2 and 11.3, respectively.

The identification results are shown in Fig. 11.4. As can be seen from Fig. 11.4, as an unsupervised machine learning recognition method, the accuracy of hand written recognition is not high at present, which needs further improvement. For example, in

Fig. 11.2 Example of Arabic number 5 written in the left hand

Fig. 11.3 Example of Arabic number 5 written in the right hand

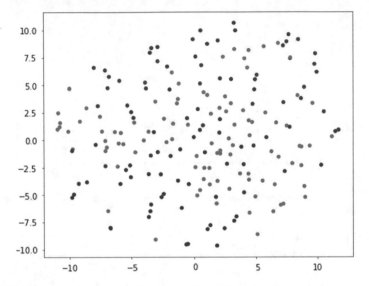

Fig. 11.4 Left and right hand recognition results of Arabic number 5

the front, only 0°, 45°, 90°, and 135° corresponding to horizontal and vertical strokes are extracted. If it is a little more elaborate, it can be more intensive between 0° and 45°, that is, increase to 5°, 10°, 15°, 20°, 25°, 30°, 35°, and 40°. Similarly, feature extraction is more intensive between 45° and 90°, between 90° and 135°, between 135° and 180°.

11.3   Thinking over Oracle Bone Inscriptions

However, the interest in unsupervised machine recognition that inspires our adversary writing stems from the question of whether oracle writes with the left or the right hand. Because our main purpose is to study whether ancient humans used to use left hand or right hand with the help of the ability of deep learning to distinguish features.

In modern society, right handedness is an obvious feature of human beings. Anthropological statistics have found that the number of right-handed people exceeds that of left-handed people, and the proportion is 9:1. In daily life, we know that the forelegs of cats and dogs are not divided into left and right hands, and their use of forelegs is balanced. But studies have shown that upright gorillas have a preference for left and right hands. Brain anatomy research found that human left brain is mainly responsible for logical thinking, language ability, and the processing of words. Right brain is responsible for creativity, intuition, and expert in images.

Most people with left handedness have a right brain advantage, while most people with right handedness have a left brain advantage. Physiologist's viewpoints come down to, (1) Theory of brain structure, the structure of human brain determines that most people use their right hand; (2) Theory of social imitation, because most people use the right hand, and the new born also use the right hand. Geneticists also gave their own explanation, and they found that handedness was related to heredity. According to the survey of American scholars, among 2177 first-year college students, left handedness is about 5%. The children whose father is left-handed accounted for 9%, the children whose mother is left-handed accounted for 13%, and the children whose parents are both left-handed accounted for 46%.

However, how long has the human right hand been used to? David Phil, a professor of anthropology at the University of Kansas, studied ancient tools and human remains to find out. He used the dents in the fossil incisors as evidence to prove that 500,000 years ago, humans were right-handed. Moreover, in collaboration with his colleagues in Croatia, Italy, and Spain, his research shows that the special traces on these teeth fossils are related to the right or left handedness characteristics of prehistoric human individuals. Phil said that the trace on the fossil teeth is directly caused by people's continuous use of tools with their right or left hands. When people do a certain action, their hands and mouth will be very close. When the stone tools were pulled, they accidentally crossed their lips, leaving these marks. The oldest fossilized teeth he used were found in a cave more than 500,000 years old, known as Sima de Los Huesos (SH), in Spain.

Oracle bone inscriptions were written in the Shang Dynasty (about the seventeenth century BC to the eleventh century BC), with a history of about 3500 years. Oracle was discovered by Wang Yirong in 1899. It has been 120 years. Oracle bone inscriptions are the earliest mature writing system discovered in China, the source of Chinese characters, and the root of Chinese excellent tradition. Oracle, an ancient Chinese character, is the source of the development of Chinese characters and the symbol of Chinese civilization for thousands of years. It has lofty symbolic significance and is often engraved or written on tortoise shells and animal bones. Oracle bone inscriptions usually appear in the form of pictograms in the publications. Oracle bone inscriptions were first unearthed in Anyang City, Henan Province. Most of them are records of divination by the royal family of yin and Shang Dynasties, most of which are about weather, agriculture, sacrifice, hunting, and musical instruments. Oracle bone inscriptions have begun to have a character system, and the modern Chinese character system has gradually evolved on the basis of oracle bone inscriptions. Oracle is carrying the code of Chinese civilization gene, is the source of Chinese character development, and is an indispensable part of the five thousand years of Chinese civilization.

Obviously, English is very suitable for the right-handed people to write, because if you use a sharp pen, writing with your left hand will break the paper, so the left-handed people who use English usually turn their wrists over to write. The invention

Fig. 11.5 Examples of ancient Chinese written with a brush

of Chinese is the same. Chinese is also very suitable for right-handed people to write. Figure 11.5 shows an example of ancient Chinese written with a brush.

Therefore, we have a strong interest in the recognition of oracle's left and right hands. However, the oracle bone itself is not big, and the characters engraved on it are also very small, with the same depth and even thickness, as shown in Fig. 11.6.

For such a small character on the oracle bone, we first guess that it should be a one handed carving. Because if you use both hands to chisel, one hand to hammer, one hand to knife, not only clumsy but also easy to chisel precious oracle bones. Oracle used to be rare, so novices must be proficient in other places before they can engrave on the tortoise shell. In addition, when using a hammer, you must use a very small hammer, and the other hand may block words, which is inconvenient. And if the hammer is used to knock, it should be a straight line, and there is no arc. Obviously, it can be seen from Fig. 11.6 that oracle was not carved with a hammer and a knife in one hand. However, only from the depth of the words engraved, it cannot be seen whether it is left hand or right hand engraved. According to the current research results of oracle experts, the seal cutting knives used for oracle inscriptions are mainly flat, oblique, and double-edged, which are generally made of jade or bronze. The carving of oracle bones is made by one hand holding a knife. (https://www.360doc.com/content/17/1001/09/37583807_691515968.shtm, https://

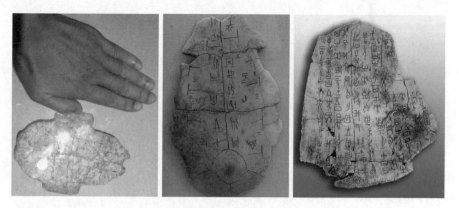

Fig. 11.6 Size of oracle bone inscriptions and examples of characters

baijiahao.baidu.com/s?id=1620567222705348982&wfr=spider&for=pc). If we say that those who wrote brush characters in ancient times were "following the trend" and Chinese characters were suitable for right-handed writing, the habit of writing with a brush in the right hand should also be corrected. But oracle bone inscriptions are not necessarily, because there is no need to "follow the trend" in that era, and both hands may be mainly used for production work.

However, with the naked eye, it seems that we cannot give whether a certain oracle type was carved with the left hand or the right hand of ancient humans. Can we find out whether the author of these oracle bone inscriptions is right-handed or left-handed through the ability of in-depth learning to accurately identify the characteristics?

Chapter 12
Social Computing Application in Online Crowd Behavior and Psychology

This chapter will introduce readers to online group behavior and psychology, an application case using online social network rumor detection. In the last section of this chapter, we will briefly introduce other examples of online group behavior and psychology.

12.1 Introduction

12.1.1 Online Social Network Rumors and Traditional Rumors

In the big data environment, "information overload" and "information pollution" have become one of the main bottlenecks in the development of many applications of online social networks, and have caused a series of problems such as "information anxiety" and "information lost" of users. Therefore, online social network rumor detection is an effective means to improve the quality of online social network information ecological environment and enhance user experience. At present, the theoretical research on online social network (OSN) rumors is mostly based on network rumors, and combined with active OSN platforms such as microblog and Twitter, a series of practical research is carried out.

For the definition of network rumor, most scholars mostly from the perspective of media theory, that is to say, network rumor is only a media from the traditional interpersonal media or mass media to the network media. This case defines online social network rumor as: online social network rumor refers to the interpretation or elaboration that online social network has played a key role in a certain stage or the whole process of its generation, dissemination and influence, the content is not verified, and has caused certain influence of social public opinion.

© Tsinghua University Press 2020
X. Liang, *Social Computing with Artificial Intelligence*,
https://doi.org/10.1007/978-981-15-7760-4_12

Table 12.1 Comparisons between online social network rumors and traditional rumors

Category	Traditional rumor	Online social network rumor
Rumor content characteristics	The facts and data are insufficient and the content is easy to change	"True rumors", content and form are more fixed and can be recreated
Characteristics of communication channels	Spatiotemporal constraints, no feedback, and difficult traceability	Break through time and space constraints, feedback, traceability, cost reduction
Audience group characteristics	Passivity	Duality, individuality, and virtuality
Characteristics of communication impact	Scope limitation	Wide and difficult to control

OSN rumors inherit the dual characteristics of OSN and rumors, but also have personalized characteristics different from traditional rumors. Next, this paper compares traditional rumors with OSN rumors based on the 5W elements of information dissemination, namely who, what, in which channel, to whom, with what effect, as shown in Table 12.1 [1].

12.1.2   Key Issues in OSN Rumor Detection

The 3W which needs to be clarified in the description of rumor detection problem is the detection target (what), the detection object (whop), and the detection time (when). According to the different arrangement and combination of the above three elements, we form the research problems suitable for different scenarios. In order to facilitate the formal expression of OSN rumor detection, the general entities and their descriptions are given in Table 12.2.

Table 12.2 Symbolic description

Expression	Description
$G = (V, E)$	G refers to OSN, V refers to nodes (users) in OSN, E refers to edges (relationships) in OSN
S	Information flow in online social networks
$e_i \in E\{e_1, e_2, …, e_n\}$	Popular topic event set in OSN
$d_i \in D\{d_1, d_2, …, d_n\}$	Single-text information set in OSN
$u_i \in U\{u_1, u_2, …, u_n\}$	User set in OSN
$f_i \in F\{f_1, f_2, …, f_n\}$	Feature set of the object to be detected
$t_i \in T\{t_1, t_2, …, t_n\}$	Propagation time series
$r_i \in R\{r_1, r_2, …, r_n\}$	The set of rumor results detected

(1) **Problem definition based on target attribute**

At present, the detection targets of rumors can be roughly divided into three categories: detection of rumors, detection of false rumors, and detection of false information.

1. OSN rumor detection problem

The first category of research objectives is also the choice of most researchers [2]. They believe that rumors, although not uniformly defined, are characterized by "unproven" (the results are true, false or still uncertain). Therefore, their detection target is the special elaboration of those facts that have not yet been determined. At this time, rumors are defined as information whose authenticity is uncertain, and information which has not been confirmed by official news is regarded as a candidate for rumors to be detected.

2. OSN error rumor detection

The second kind of research goal is that the false rumor information is more harmful than the true information disseminated into rumors [3, 4]. At this time, researchers often use the words "false rumor," "misinformation," and "disinformation" to distinguish them from the word "rumor," so as to refer specifically to those rumors which are finally proved to be false.

3. Detection of false information in OSN

The third kind of research goal is not to classify rumors according to their final confirmation results [5], but to equate rumors with false information. At the same time, there is no distinction between the words such as "rumor," "misinformation," and "disinformation." This kind of research directly transforms rumor detection into false information, error information recognition, or information reliability ranking problem.

To sum up, there are several main problems in the determination of research objectives. First, the detection of rumors needs to clarify the concept and essence of rumors, strictly distinguish the difference between rumors and false information, and analyze the difference between rumors detection and false information recognition and information reliability recognition in detail. Secondly, although the false information in rumors is more destructive, the harmfulness of the information which is ultimately correct and still cannot prove its authenticity cannot be underestimated. The purpose of refuting rumors is to seek the truth so as to prevent illegal information from disturbing the public's vision and thus to build a good information ecological environment. Thirdly, among the three categories mentioned above, researchers compare rumors/false information and news as a pair of concepts. However, the credibility of the current network media is gradually weakening, in order to win the public's attention, the news media often shorten the link of news confirmation, and false reports can be seen everywhere. Therefore, the reliability of official news is also questionable.

(2) **Question definition based on object attribute**

A rumor is a statement or statement that people are interested in or feel important and that the facts are not yet certain. It usually arises from one or more rumor makers, and is widely circulated among the crowd and finally forms a rumor information group about the statement or elaboration. It can be seen that the detection objects of OSN rumors can be divided into fine-grained detection objects based on single-text information and coarse-grained detection objects based on multi-text information from the detection granularity, and also include rumor source detection objects from the information source angle.

1. Rumor detection in single-text information OSN

The detection of OSN rumors with single-text information is similar to the detection of false or spam information in OSN. The detection target is simple and clear, and the feature is easy to extract. However, the drawbacks of this kind of problem are also obvious: firstly, in OSN, the amount of information contained in a message is extremely limited. Therefore, in the design of the detection model, we still need to consider the valuable signal features in the text cluster, like the overall propagation characteristics, network structure characteristics, etc., in order to make up for the limited signal features in a single text. In addition, the detection of a single message usually does not meet the needs of the real scene. The object of rumor refuting is usually assertion in a topic statement. That is, a judgment statement around an event or person. Users are also concerned with the assertion of the statement, not with a single message. Once confirmed as a rumor, the information consistent with this statement will be regarded as a rumor by the user and will not be accepted.

2. Rumor detection in multi-text information OSN

OSN rumor detection based on multi-text information regards the object as a topic or event containing multi-text information [6]. Usually, after the user of the information source publishes statement information, the relevant user carries out a series of forwarding or comments, and the above text information sets are all around the original statement information. OSN rumor detection model extracts the content features, user features, network features, and propagation features contained in the transmitted text information, and then establishes feature vectors to classify and recognize them.

OSN rumor detection based on multi-text information can generally be divided into two kinds in specific applications: daily rumors for ordinary events and special rumors for emergencies. Generally speaking, after emergencies (such as earthquakes, tsunamis), rumors about different topics will be generated around the event. Therefore, OSN rumor detection for multi-text information of emergencies mostly takes the emergencies as a whole, and then identifies all kinds of rumors in the event. However, this kind of detection often relies on the topic information of the event and does not have the universality of detection.

3. Rumor source detection

Rumor source detection refers to the abstraction of, OSN, rumor network into a certain information dissemination model (such as infectious disease dissemination model SI, SIS, SIR), and then according to the characteristics of rumor dissemination to find the original node of rumor dissemination, that is, rumor source. It is generally believed that rumors are always generated from a small number of groups, so the detection of rumors can be divided into single-source detection and multi-source detection. In the specific research, in order to simplify the detection problem, we usually only consider the single rumor source detection problem.

(3) **Question definition based on time attribute**

According to the time attribute of OSN rumor detection, it can be divided into three core issues: retrospective OSN rumor detection, early OSN rumor detection and real-time OSN rumor detection.

1. Retrospective OSN rumor detection

Retrospective OSN rumor detection is mainly aimed at the old rumors on OSN or the historical and static rumor data set during and after the rumor outbreak period. That is to say, the training and testing of the model are based on the historical data before the rumor outbreak period or the end period, which is also the most widely studied model at present. The advantage of retrospective detection is that there are more trainable data sets, so that more signal features can be extracted, especially the number of propagation features (such as forwarding number, comment number). However, the retrospective detection method is not suitable for real OSN scenarios. When an emergency breaks out, it is urgent to identify rumors as early as possible and suppress their spread in order to minimize the negative impact of rumors.

2. Early OSN rumor detection

Early OSN rumor detection refers to the detection of rumors when they are still in latency. Early OSN rumor detection is generally based on event level, that is, multi-text information detection. With the passage of time, the text information contained in the event description is increasing, so at least one text information is included in the event description during detection. At the same time, the purpose of early detection is to shorten the time difference between the first text information elaborated on the event and the detection time as far as possible.

3. Real-time OSN rumor detection

Real-time detection is based on rumor real-time data stream detection, usually for single-text information, that is, once the information is released, it begins to detect whether the information is a rumor or rumor candidate.

In summary, the time attribute characteristics of rumor detection can be summarized as follows: First, the fundamental difference between retrospective detection, early detection, and real-time detection is whether to introduce time parameters. Because rumors have strong timeliness, early detection and real-time detection are more in line with the actual needs and can effectively reduce the possible adverse effects of rumors. Secondly, compared with retrospective detection, early detection and real-time detection have fewer features of rumor signals available, so they are more challenging. Thirdly, as far as the accuracy of rumor detection is concerned, it is generally difficult to achieve 100% accurate detection. In contrast, the retrospective detection method has the best detection accuracy because it can acquire the most signal features, generally reaching about 90% [7]; the second is early detection, and the third is real-time detection. Generally speaking, there is a direct relationship between the accuracy of rumor detection and the attributes of rumor detection time. The earlier the detection time is, the greater the detection effectiveness (i.e., suppressing the possible harm of rumors as soon as possible), but the detection accuracy is relatively low. On the contrary, the later the detection time is, the smaller the detection utility is, but the detection accuracy is constantly improving.

12.1.3 OSN Rumor Detection Open Data Resources

At present, there are few open data resources available for OSN rumor detection research. Meanwhile, the research platform mainly focuses on Twitter and Sina Weibo. According to the different research objects, it can be divided into two types: message-based and event-based. As shown in Table 12.3, it is a common data set for

Table 12.3 Open OSN rumor data set

Platform	Microblog number	Topic/number of messages (r/nr)	User	Data link
Twitter	7507	2695/4812	7507	https://ndownl oader.figshare.com/ files/4988998
Twitter	5802	1972/3830	–	https://www.pheme. eu/software-downlo ads/
Sina microblog	5137	2601/2536	5137	https://adapt.seiee. sjtu.edu.cn/~kzhu/ rumor/
Twitter/Sina microblog	1.12 million 38.05 million	498/499 thousand 2313/2351 million	490 thousand 2.74 million	https://alt.qcri.org/ ~wgao/data/rum dect.zip
Twitter		51/60		https://dataverse.har vard.edu/dataset. xhtml?persistentId= doi%3A10.7910% 2FDVN%2FB FGAVZ
Sina microblog	110 thousand	50/274	88 thousand	https://www.dro pbox.com/sh/9lm y4veobd2oknk/ AABEcn77P RHwK%20JcNJitm 7d0Ma?dl=0

current OSN rumors. Message-based data sets only collect and annotate single-text data (rumors or non-rumors); event-based data sets collect multi-text contained in topics and annotate topics.

Message-based public data sets include PHEME (the European Commission's three-year research project fund), Zubiaga et al. [8] public data sets, and Wu's [4] experimental data collected for Twitter and Sina Weibo, respectively. The amount of data is not large (less than 10,000), and the ratio of rumors to non-rumors Weibo is approximately 1:1. The typical representative of event-based public data sets is Ma et al. [9], which collect large-scale data sets for Twine: and Sina Weibo, including 1.12 million tweets and 3.805 million microblogs, including 498 Twitter rumor topics and 2313 microblog rumor topics, as well as 490,000 Twitter users and 2.74 million microblog users. By contrast, Kwon et al. [10] collect relatively few rumor topic data sets for Twitter and Jin, etc., for Sina Weibo.

For Sina Weibo rumor experiment, the well-known and effective open experimental data set also includes the Chinese rumor database (https://rumor.thunlp.org) [11] established by Sun and Liu, Laboratories of Natural Language Processing and Social and Humanistic Computing, Tsinghua University. There are three main sources of data: data capture of Chinese social media rumors, automatic identification of Chinese social media rumors, and user submission. The downloaded rumor data set spans from May 2012 to June 2016, with a total of 3038 rumor messages, which is still being updated.

12.2 Content Classification Based on Machine Learning Rumor Detection

Content rumor detection as a machine learning classification problem for single or multiple text is in information flow S. In machine learning, supervised or semi-supervised learning algorithms are often used for classification problems, that is, the task of training prediction functions from labeled training data. The training data are composed of labeled OSN rumors (i.e., data set input objects matching expected output objects), and the feature vectors are extracted. Then, the data are analyzed by relevant algorithms, and the predictive model functions that can be used to discriminate the new input objects are inferred. Finally, the expected predictive results are obtained [12].

In previous studies, the differences of OSN rumor detection classification models mainly include the following key points: input text granularity, data set feature selection, classification algorithm selection and classification output object type. (1) Input text granularity is divided into coarse-grained multi-text input object (event-based) and fine-grained single-text input object (message-based). (2) Feature selection of data sets, which is to find out the difference between rumors and normal information, and then construct the feature vector of classifier input. At present, the common feature vector selection includes content feature, user feature, network feature, propagation feature, and their combination feature. (3) Selection of classification algorithms, and the common classification models can be divided into two categories: rule-based classification methods (e.g., decision tree, association rules, and rough sets) and statistical-based classification methods (e.g., naive Bayesian, support vector machine). (4) Classify the output objects. For the binary classification problem, the combination of output objects includes false information/real information, rumor/non-rumor, false rumor/non-false rumor, etc. In addition, there are outputs of multi-classification results.

In each step of the classification model, the accuracy and validity of the classifier largely depend on the selection of input features. Specifically, it mainly includes the methods of text features and non-text features.

(1) Method of text feature

Text features refer to the dominant features of message text content which are mainly grammatical, implicit features which are mainly semantic and emotional, as well as the new features which show over time. The content features of microblog message sets contained in microblog messages or events are generally better than other user-based, network-based, or propagation-based features [13, 14].

1.   Grammatical dominant features

Grammar-based dominant feature analysis mainly includes the lexical features, symbolic features, and simple emotional features [15] of the content of the message text, as shown in the table below. Among them, word features and symbol features are mainly analyzed by simple manual statistics (such as whether to include emoticons and their number). Simple emotional feature analysis is to compare the number of positive and negative emotional words in the text information, and ultimately determine the emotional orientation of the text. Among them, the emotional vocabulary collection comes from WordNet (English domain) [16] and HowNet (Chinese domain) [17].

Machine learning detection method based on dominant text feature classification is a common method used in early OSN rumor detection. For example, Castillo, a pioneering researcher in rumor detection, subdivides the text features of [18] into string length, number of words, whether it contains emoticons (e.g., smiles and frowns), personal pronouns (first person, second person, third person), capital letters, publishing time, number of positive words, number of negative words, whether it contains labels, whether it contains links, etc. In addition, Ratkiewicz et al. [19] established Truthysystem to detect rumors by using tags, links, and questions contained in political rumors on Twitter as text features (see Table 12.4).

It can be seen from the above that the early OSN rumor recognition based on text features is mainly characterized by superficial dominance. This method is simple and convenient, but only suitable for the detection of small-scale data and specific topics of rumors (such as political rumors, food safety rumors). Because a particular type of rumor topic is more likely to show its unique dominant text characteristics. However, the accuracy and universality of the whole classification model are still unsatisfactory.

Table 12.4 Classification of dominant features based on grammar

Feature class	Concrete content
Word features	String length, part of speech, upper and lower case letters
Symbol characteristics	Punctuation, emoticons
Simple emotional characteristics	Number of positive and negative emotional words
Meme feature	Does it contain tags, links?

Table 12.5 Semantic-based hidden feature analysis method

Feature class	Existing method
Deep semantic features	Word bag, neural network
Multi-granularity emotional characteristics	Sparse addition generation, word vector, and emotion classifier
Inter-message association characteristics	Semantic similarity computing

2. The implicit characteristics of semantics

Semantic-based implicit feature analysis mainly aims at feature extraction or abstract representation of message text at deep semantic level, so as to obtain latent semantic features, emotional features, interrelation features, and other text features of message text, as shown in Table 12.5.

(a) **Deep-seated semantic feature analysis**

For deep-level semantic feature analysis, the existing methods usually ignore grammar and word order features, and abstract message text through semantic representation learning to obtain its deep-level semantic features. Common methods include word bag model and neural network model.

The basic idea of the bag-of-words model is to assume that for a text, the word order, grammar, and syntactic features are ignored, and it is only regarded as a set of words (and each word in the text is independent); and then each document is represented as a word vector in the vector space, so that documents with similar content have similar vectors. The single word or multiple continuous words in the set of words can also be expressed as the bag features of the text, in which the unigram, bigram, and trigram are the commonly used bag features. Word bag feature is a very effective feature in view mining and emotional analysis of text. For example, the researchers used the word bag model and the neural network language model (Word2vec (100) and Word2vec (400)) to compare the two text representation models. In the test of 10,000 posts on Sina Weibo, we found that the best classification accuracy of the word bag model was more than 90%, and the best accuracy of the neural network language model was more than 60%.

Neural network has a great reputation in the field of machine learning, but because it is easy to over-fit and its parameter training speed is slow, and the traditional artificial neural network is a shallow structure, which is far from the expected artificial intelligence, it gradually fades out of people's vision. However, with the improvement of processing speed and storage capacity of computer, the realization of deep neural network is becoming possible gradually, such as the different forms of cyclic neural network/convolutional neural network. The deep neural network is a breakthrough in the traditional feature selection and extraction mode, so the semantic features of commonly used natural language processing are abstracted. Compared with the prior learning algorithm based on manual crawling feature, the model based on RNN has a great improvement. Through three widely used recursive units: tanh function, long-term and short-term memory, and threshold recursive unit, the hidden layer expression of learning messages achieves a higher accuracy than the existing detection methods. The deep neural network modeling method uses continuous vectors to represent text, which overcomes the problem of sparse features, and simulates the thinking mode of human brain to a certain extent, which has a high accuracy rate. However, the model has many parameters, slow learning convergence, and needs a large number of corpus to calculate.

(b) **Emotional analysis**

Emotional analysis also belongs to a method of text semantic analysis, that is, to analyze the semantic characteristics of message text from the perspective of emotional polarity. Because rumors are driven by psychological and emotional factors to a large extent in the process of dissemination, and compared with positive emotions, inciting negative emotions is more likely to accelerate the spread of rumors. Therefore, emotional characteristics, as one of the important features, have been studied by many researchers in the text features of rumor detection. The granularity of affective analysis generally includes affective analysis at text level, sentence level, and word level. In the specific research methods, similarity computing and machine learning are still the mainstream.

Compared with the emotional characteristics of the main content of microblog, the text emotional characteristics of the forwarding content and comment content of, OSN, group are also concerned. For example, based on the hypothesis that rumor-based microblog comments tend to be more negative than ordinary microblog comments on the whole, a classifier based on word frequency features is proposed to effectively identify the emotional inclination of a single comment, thus obtaining the overall emotional inclination feature value. Finally, an experimental verification based on microblog corpus shows that the proposed new features can significantly improve the classification results on the basis of existing features.

(c) **Implicit association analysis**

The research on the implicit relationship between microblog information mainly refers to the analysis methods of semantic, emotional, and situational correlation between messages and messages. Semantic similarity computation is a common method for semantic association analysis. For example, it is assumed that the observed and missing semantic spaces of words can constitute the complete semantic features of sentences, and then the latent semantic of sentences is extracted by using semantic textual similarity (STS) model to create the latent vector representation (TLV, tweet latent vector) of each microblog. STS preprocesses each short text by marking and supplementing, then deletes unused words, and uses tf-idf to empower, finally extracts latent semantics, and forms a 100-dimensional latent semantic feature vector.

For the affective feature correlation analysis and situational feature correlation analysis, simple analysis methods include the use of external auxiliary information, such as tags or links. The deeper mining method uses the way of modeling the rumor event propagation network to depict the rumors of various topics, as well as the potential relevance of each message level and within the layer.

Taking the false news "Shenzhen's Most Beautiful Girl feeds the old beggars on the street" in 2013 as an example, the false news example constructs a three-level credibility network communication model: message layer, sub-event layer, and event-layer. The three-tier trustworthiness network forms four types of network links to reflect the implicit relationship between network nodes, namely the intra-tier links (message-to-message, sub-event-to-sub-event) represented by dotted lines and the inter-tier links (message-to-sub-event, sub-event-to-event) represented by dotted lines.

The intra-layer links reflect the relationship between entities in the same level, while the inter-layer links reflect the relationship between levels and levels. For a given news event and its related microblogs, sub-events are generated by clustering algorithm, and implicit semantic information between links is captured by sub-event layer. In the experiment process, three independent models are generated based on content feature, communication feature, and user feature. At the same time, logical regression is used to synthesize the above three independent models to get a comprehensive model. Finally, the visualization results are constructed.

In summary, the analysis of message text semantic features can improve the accuracy of prediction by mining the deep semantic features, emotional features, and association features of the text, compared with the grammar-based dominant feature extraction. Represented by the neural network model, semantic abstraction and feature extraction of text in the form of vectors is a new research method of rumor detection based on in-depth learning in the future. However, the current research on text content mainly focuses on the content of the message text itself, and lacks the semantic analysis and association analysis of the commentary text, forwarded text, which is also the place that needs to be further improved in the future.

(2) Multimedia information feature method

In the spread of OSN rumors, multimedia content (such as pictures, audio and video) is more attractive and persuasive than text information alone. Sun et al. [19] pointed out that 80% of the rumors contain pictures. It can be seen that multimedia content feature is an effective complement to text content feature in OSN rumor detection. However, in the existing research, the depth of mining multimedia information content features is insufficient, most of which only use multimedia as a supplement to text features, while the research on detection of multimedia information itself as rumor media is less (especially for in-depth analysis of video or audio information). For example, only a small amount of extraction and analysis of image features uses the content label or external knowledge (such as search engines) to verify the content of the image, but does mine the content characteristics of the image itself. Sun and others mainly studied the types of rumors with inconsistent pictures and texts, and pointed out that 80% of the rumors contain pictures, and the need for pictures is limited, of which 86% of the false pictures are obtained by forwarding. In the specific research, they divided rumors into four types: fictional news, outdated news, tampered news, and inconsistent news. They also used image features of multimedia features to detect rumors of image–text mismatch. In the process of implementation, the author submits the pictures of Weibo to the external search engine to retrieve the source of the pictures. If no result is returned, the pictures and texts of Weibo are consistent. Otherwise, the recorded results will be sorted in descending order according to the credibility of the website and the release time of the original pictures. Then, we crawled the top-ranked website content, and used Jaccard coefficient to calculate the similarity between microblog content and crawling website content. If the similarity proves that the microblog is not a rumor, otherwise, it is a rumor that the pictures and texts do not match. Experiments show that compared with common benchmark features, image features have better detection results on various classifiers.

From the above analysis, multimedia content has gradually become an unavailable part of rumor dissemination process, but the current research based on multimedia features to detect rumors is very scarce, and mainly embodied in two aspects: first, although the current research takes into account the rich information contained in the external label of the picture, the effective information really used is seriously insufficient. In fact, an image uploaded by a user contains a lot of metadata information: photo title, ID, shooting time, shooting location, and the relevant personal information of the photographer [20]. Secondly, in current research, deep mining based on multimedia content is relatively rare, that is, it does not use relevant multimedia content processing technology to identify its inherent semantic features. Therefore, in the future research, we need to further fully apply multimedia tag content and multimedia semantic mining technology to rumor detection research.

(3) **User behavior feature method**

OSN rumor detection methods based on user behavior characteristics generally include four types: information publisher behavior characteristics analysis, information transmitter behavior characteristics analysis, information receiver behavior characteristics analysis, and interaction behavior characteristics analysis among users.

For the analysis of behavior characteristics of information publishers, it is usually based on the features of collecting users' recent posts, forwarding, attention, fans, and abnormal posting patterns. This is also the most commonly concerned user behavior characteristics in the existing research: the behavior characteristics of rumor publishers. For example, by identifying extreme users to detect political rumors on Twitter, it mainly includes whether the user's introduction contains extreme keywords, whether the latest 100 tweets of users contain extreme keywords, and other common user characteristics such as user fans. The limitation of this method lies in the strong dependence of setting parameters manually. Experiments show that the best rules for different data sets and rumors are different. So, it may not be possible to know which rules are best for a given new data set. Secondly, the use of extreme keywords to identify extreme users is effective in detecting political rumors in this text, but in other types of rumors, extreme users do not necessarily use extreme keywords.

Generally, the analysis of the behavior characteristics of the information transmitter is based on the specific behavior of the forwarding user, such as the attributes of the forwarding user (e.g., the number of fans, the number of concerns), and the abnormal pattern of the forwarding user. Usually, the analysis based on the behavior characteristics of the transmitter is also called the analysis based on the propagation characteristics, that is, the analysis based on the propagation structure formed by the forwarding path. For example, the researcher clearly points out the difference between rumors and non-rumors in the process of spreading: rumors are usually issued by ordinary users, then forwarded by opinion leaders, and finally forwarded by a large number of ordinary users. Non-rumors are usually released by opinion leaders and then forwarded directly to ordinary users. In the specific research, the author combines the forwarder's behavior characteristics with the information publisher's behavior characteristics and message content characteristics, and uses the hybrid SVM classifier to predict rumors.

For the analysis of information receiver's behavior characteristics, we usually extract important features from user's comment behavior to distinguish. For example, researchers focus on users' response behavior in emergencies to identify Sina Weibo rumors. The validity of feature set selection is verified by extracting user's forwarding and comment text features, and then clustering analysis. Finally, rumor detection is realized by machine learning classification technology. Experiments show that stop words, punctuation marks, and some words or symbols that express people's response to emotions have played an important role in detection.

For the analysis of interaction behavior characteristics between users, in fact, more and more user behavior feature analyses are to extract the comprehensive features of user behavior pattern and interaction pattern. For example, researchers have established rumor detection classifiers based on the characteristics of users of microblog publishers and microblog readers. Five new features are proposed, which are different from previous studies, including three behavioral characteristics based on microblog publishers: average daily number of concerns (total number of concerns/days of registration), average daily number of posts, number of people publishing similar to a specific microblog (article assumes that rumors always come from a few or even one person, while normal information may have multiple sources), and microblog based on two behavioral characteristics of blog reception: the proportion of questioning comments and the number of correcting comments point out that 14.7% of people or organizations will release rumor information in time after discovering rumors. The experimental results show that the new features are effective in rumor detection.

The spread of rumors is a two-way process, which exists in the whole ecological chain of rumor information dissemination: rumor producers, rumor transmitters (forwarders), rumor receivers, rumor decomposers (rumor dispellers), and other different users. However, in the existing studies, the main focus is on the characteristics of rumor producer's behavior. In fact, rumor transmitter's behavior and rumor receiver's behavior contain a lot of effective potential information for rumor detection [21]. For example, the behavioral characteristics of rumor forwarders mentioned above, the responding behavioral characteristics of rumor receivers, and the interactive behavioral characteristics of the two. Therefore, in the future research, the research on user behavior in rumor communication needs to further integrate the behavior characteristics and interaction characteristics of producers, transmitters, and receivers.

(4)   Method of time attribute characteristics

With the passage of time, rumor propagation mode and effective detection signal characteristics may change. Most of the existing studies are based on an arbitrarily set single observation window, and the results obtained from a fixed observation point are often difficult to express the general rumor propagation mode. In this regard, the researcher combines user, structure, language, and time characteristics to analyze the performance level of rumor classification detection from the first three days to the last two months in different time windows. Statistical analysis shows that structural and temporal features are better distinguished in long-term windows, but not in the initial stage of transmission. In contrast, user and linguistic features are easily acquired in the initial stage of rumor propagation, and they are also good indicators of rumor detection in the initial stage of rumor propagation. Therefore, the importance of time attribute features in rumor detection is obvious. In fact, dynamic prediction is one of the most important challenges for OSN rumor detection. In practical research, there are two main approaches to the dynamic nature of OSN rumor detection: first, real-time flow data detection, and second, time-varying information, user or network structure model detection.

For path 1, rigorously speaking, rumor detection based on dynamic streaming data is less likely to adopt content-based classification machine learning method, because this method requires a certain amount of initial data training to start the test. For example, by adding two kinds of new features: information source reliability and user attitude, we can obtain different amounts of real-time data (5, 100 and 400) and compare the classification accuracy after different time delays (1 h later, 12 h later, 72 h later). And it is found that over time, the more initial data available, the higher the classification accuracy.

For path 2, some studies pointed out the importance of time attribute in rumor spreading process, and proposed a time series fitting model of the number of tweets changing with time. Experiments show that better detection results can be obtained. Thereafter, Ma et al. [22] expanded the feature set with time on the basis of Kwon's research. Dynamic time series was used to observe the changes of social scene features over time, including text features, user features, and propagation features. Finally, the SVM classifier based on dynamic time series is used to obtain better detection results on the Twitter data set and Sina Weibo data set.

Because rumor detection often depends on external factors such as topic, network model, and so on, it is difficult to achieve a truly universal model. Therefore, when using machine learning classification method, in order to effectively identify rumors, a certain amount of training set is needed to train the model, which has become one of the drawbacks of real-time rumor detection. Thus, rumor detection based on machine classification has the best effect in retrospective detection. The improved model can also be used for early detection by adding certain time attributes, but it is still difficult to achieve real-time detection. The detection of real-time rumors also depends on the content-based comparison method discussed below.

(5)   **Method of geographical location characteristics**

Generally, there are three types of geographic location characteristics in the process of rumor dissemination: the location of communication network, the location of rumor events, and the location of information publisher (or the location of information disseminator). In OSN networks, in different regions of the world (such as Twitter and Sina Weibo), the characteristics of message transmission and people's information behavior are different, so rumor propagation network space also affects detection performance. In this regard, the researchers selected nine experimental data sets of rumors from different parts of the world to construct a topic-independent rumor classifier, but the classification effect is not satisfactory. Experiments show that most of the features of rumor recognition are based on specific environment, and the location of the event (different countries, different social groups, different cultural characteristics) plays a key role. Therefore, in the future rumor detection model, it should be tested in different network environments and different types of data sets to ensure the realistic availability of the model.

At the same time, the distance from the place where the rumor happened will also affect the spread of rumors. For example, two new non-text features are extracted based on the location of event occurrence and the message publishing client. Based on the location of the event, it refers to the location of the event mentioned by the message, which can be divided into two types: domestic and foreign. Client-based features refer to the client used by users when publishing messages, which can be divided into smart client and non-smart client. It is found that rumor information is more likely to be published on webpages, and the locations of the events involved in rumor information are mostly in China. Finally, combining location features and client features with three types of features based on message, account and communication (19 in total) to train SVM classifiers of RBF kernel functions to identify rumors, which can achieve good results.

With the development of Web 2.0 and location-aware technology, social networks allow people to add additional time stamps and geographical locations as tags, which makes it easier to share and search uploaded files or published content based on temporal and spatial information. For example, more than 40 million photos on Flickr and more than 4 million photos on Panoramio are geographically tagged.

These publicly accessible photos not only contain rich geographic information but also convey the feelings and opinions of people of various demographic backgrounds. These tags are highly personalized. In addition to the common location of geographical names (such as Beijing, Great Wall), they may also contain date, weather, camera parameters, and even the user's personal mood. In rumor detection, the current location of the information publisher or the specific location of the message, as well as the distance between the location of the publisher and the location of the rumor event are important information worthy of attention.

Geographical location features provide a new and effective signal feature for OSN rumor detection. Especially for rumors caused by sudden natural disasters or major social accidents, the location of information publishers and events become important features of rumor recognition. At present, the location information mining is mostly limited to the location information of external tags based on LBS technology. In fact, the image attribute information uploaded by users will also contain the location information of the message published. However, there is no research to make full use of this effective information, which further illustrates the importance of multimedia information for rumor detection.

12.3 Network Structure Based Rumor Source Detection

The ultimate goal of rumor detection is to block the spread of unconfirmed information in a timely and effective manner and to prevent its possible adverse social impact. Among them, the identification and control of rumor source are very important. It can effectively find the root of rumor spread and control its further spread. Rumor detection based on network structure focuses on the detection of rumor sources, and belongs to the problem of information source inference. It abstractly describes the social network topology structure in the form of graphs, and abstracts the propagation model of information in the social network. Then, it constructs the node estimator of rumor sources according to the snapshot of infection propagation sub-graph, so as to maximize the accuracy of the estimation.

In the rumor source detection model based on network structure, two basic models need to be clarified first are rumor information dissemination model and information network model. For rumor information dissemination model, besides SI, SIS, and SIR models mentioned above, there are also common independent cascade models (independent cascade model and linear threshold model). For information network model, representative models mainly include rule network, random network, small-world network, scale-free network, and real network, among which rule network is the preferred information network model for researchers.

On the specific identification object, rumor source detection is divided into single rumor source detection and multi-rumor source detection. Single rumor source detection is a simplified form of rumor source detection, and it is also the focus of researchers. The detection of multiple rumor sources can also be transformed into the detection of multiple single rumor sources. For example, researchers suggested that in practical applications, rumors always erupt from a cluster of nodes and are sent out by multiple rumor sources. The primary rumor source is a kind of node of the main rumor source node. Because of its great influence or abundant connectivity, it becomes the most important source node. Therefore, the multi-source detection problem can be transformed into the detection of the primary rumor source node. In terms of specific detection methods, rumor source detection based on network structure is mainly divided into detection methods based on spread sub-graph snapshot and detection methods based on deployment observation points.

(1) Detection method based on propagating sub-graph snapshot

The detection method based on propagating sub-graph snapshots is to obtain the status sub-graph of whether all or part of the nodes receive rumor messages (i.e., whether they become infected nodes) one or more times, then estimate the measurement characteristics of a network's topological attributes, and then deduce the nodes that are most likely to become rumor sources in the network. This is also the most common method about rumor source detection at present. Among them, the rumor-centered C rumor centrality proposed by Shah and Zaman [23] is the most common measure of network topological attributes of propagating sub-graphs, i.e., source point estimator based on maximum likelihood estimation of combinatorial number. In addition, other transmission sub-graph detection methods, such as critical edge probability of transmission and Simulation of infection path, are also included.

For the centrality of rumors, the groundbreaking research results began with the work of Shah and Zaman [24]. They assume that each node has the same probability as the source of rumors, and then calculate the number of paths from the current node to other infected nodes and the probability of the occurrence of paths according to the propagation sub-graph of infected nodes observed in the network, so as to establish a likelihood function. Finally, the maximum likelihood estimates of each node are calculated separately. The node with the largest likelihood estimates is the information source of the network. In this process, they proposed a well-known topological attribute metric for rumor source detection: rumor centrality. Rumor centrality is a "graph fraction" function, which takes network $G = (V, E)$ (where V is a countable infinite set of nodes and E is the set of edges of the form (i, j) for some i and j in V) as input, and assumes that when each node is a source node, the number of path sequences from the source node to the last infected node is the score value of each node, i.e., rumor centrality, where V is a countable infinite set of nodes and E is the set of edges of the form (i, j) for some i and j in V. The node with the highest score is the source of rumors, also known as the rumor center.

In addition to the method of combining rumor centrality with maximum likelihood estimation to locate rumor sources, other methods, such as graph-centrality measurement methods, have also been studied, such as the concept of distance centrality, which considers that in the sub-graph of information dissemination, the smallest sum of distances to other nodes is the information source of the network. Researchers put forward the concept of accessibility, which locates the node with the greatest accessibility as the information source of the network. In addition, the researchers defined the probability of rumor spreading boundary. In the snapshot of infection spreading sub-graph, the source of rumor was located by finding the node with the highest probability of spreading boundary. However, this kind of detection method based on graph centrality measurement often needs complete propagation sub-graph, which is difficult to implement in time-varying social networks. At the same time, this method also has some limitations in multi-source detection, which is not conducive to meet the actual needs.

(2)   **Detection method based on deployment node**

In the previous section, the detection method based on snapshots of propagating sub-graphs has mentioned that because of the large-scale and complexity of OSN network, it is difficult to obtain the true infection status of nodes in the propagating graph. At the same time, in the OSN network, because the importance of each node is different, if all the nodes in the network are studied, it will inevitably increase the unnecessary computational overhead. Therefore, for a given network, without knowing the status of network node infection and the relationship between nodes, a suitable number of nodes and important locations can be selected as the observation points of the whole network for research. Deployment node-based detection method is exactly based on this idea. A small number of observer nodes, or monitor nodes, are deployed in social networks to record the time and direction when they first receive messages sent by neighbor nodes, and then the source of rumors in the current network is inferred by statistical calculation.

The detection method based on deployed nodes mainly includes three steps in the specific operation process: deploying observation points, estimating information sources, and finally taking the candidate source with the largest estimated value as the rumor source of the current network. This method begins with the inference of information sources [25]. By sparsely deploying a small number of observation points in the network, the dissemination information recorded by observation points can be obtained, and then the probability that each node in the network is a real information source can be calculated to realize the location of information sources. In the practical application of rumor source detection, some researchers [12] compared the observation point selection methods of various centrality measures, and divided the observation points into positive observer and negative observer, then calculated the reachability and distance of each node to these two kinds of observation points, respectively, and then formed two indicators: greedy source set size (GSSS) and maximal distance of greedy information propagation (MDGIP). Finally, combining these two indicators, the rumor source is used to locate the source of the rumor. And some researchers [26] used independent cascade model to describe information dissemination in social networks, and define rumor quantifier, which is a probability-based value used to arrange nodes to become rumor sources. At this point, the author hypothesizes that cascades are more likely to spread from rumor sources to positive observation points (i.e., the infection points where messages are received), rather than to negative observation points. Experiments show that when there are a reasonable number of observation points, the author can effectively identify the source of rumors by calculating the rumor quantifier of each node through the scalable RSD algorithm.

From the above review, the method based on deployment of observation points only needs to obtain a small amount of communication information feedback from observation points, and does not need to peep into the whole network transmission state, so it is feasible in practical application. However, the accuracy and overhead of this method depend on the location and number of observation points deployed in the network. In this regard, the consensus reached by researchers is that the more influential nodes in the network have on information dissemination, the more likely they will receive rumor information in the process of information dissemination, and the more effective the recorded rumor dissemination information will be. In current research, centrality is an effective measure of the importance of network nodes, including degree centrality, betweenness centrality, and closeness centrality. Degree centrality of a node refers to the number of neighbor nodes of the node, which mainly describes the direct impact of the node in the network. Close centrality refers to the inverse ratio of the distance between the node and other nodes in the network. The larger the value, the closer the node is to other nodes. Median centrality refers to the number of times that the shortest path between two nodes in a network passes through the node. It mainly describes the importance of nodes in the flow of network information.

On the whole, the detection methods based on propagation sub-graph snapshots and deployment observation points have different advantages and disadvantages. Although the detection method based on spread sub-graph snapshot can effectively locate the source of rumors, it often needs a large number of dynamic infected network topology state diagrams (such as infection status of the whole network, observed infection network snapshots, observable probability of node infection status) in the process of rumor infection, and this method is often based on static network topology diagram, the timeliness of detection and the dynamics are poor. Therefore, it is difficult to apply the large-scale and dynamic social network in practice. On the other hand, the detection method based on attributes measurement of deployed observation points does not need to grasp the infection status of the whole network, and only needs sparsely deployed observation points to obtain feedback information. Therefore, the cost of computing network resources is small, but the rational deployment of the number and location of observation points is still the core challenge in current research.

References

1. Lasswell H (2013) The structure and function of communication in society. Communication University of China Press
2. Qin Y, Wurzer D, Lavrenko V (2018) Spotting rumors via novelty detection. arXiv preprint arXiv:1611.06322
3. Chen W, Yeo C, Lau C (2016) Behavior deviation: an anomaly detection view of rumor preemption. In: Proceedings of the 2016 IEEE 7th annual information technology, electronics and mobile communication conference, Vancouver, pp 1–7

4. Wu K, Yang S, Zhu K (2015) False rumors detection on Sina Weibo by propagation structures. In: Proceedings of the 2015 IEEE 31st international conference on data engineering, Seoul, pp 651–662
5. Zhang Q (2015) Exploiting the topology property of social network for rumor detection. In: Proceedings of the 12th international joint conference on computer science and software engineering. Songkhla, pp 41–46
6. Liu X, Nourbakhsh A, Li Q (2015) Real-time rumor debunking on twitter. In: Proceedings of the 24th ACM international conference on information and knowledge management, Melbourne, pp 1867–1870
7. Yang H, Zhong J, Ha D (2016) Rumor propagation detection system in social network services. In: Proceedings of international conference on computational social networks, Ho Chi Minh City, pp 86–98
8. Zubiaga A, Liakata M, Procter R (2016) Analysing how people orient to and spread rumours in social media by looking at conversational threads. PLoS ONE 11(3):1–34
9. Kwon S, Cha M, Jung K (2017) Rumor detection over varying time windows. PLoS ONE 12(1):e0168344
10. Jin Z, Cao J, Jiang Y (2014) News credibility evaluation on microblog with a hierarchical propagation model. In: Proceedings of the 2014 IEEE international conference on data mining, Shenzhen, pp 230–239
11. Dayani R, Chhabra N, Kadian T (2015) Rumor detection in twitter: an analysis in retrospect. In: Proceedings of the 2015 IEEE international conference on advanced networks and telecommunications systems, Kolkata, pp 1–3
12. Seo E, Mohapatra P, Abdelzaher T (2012) Identifying rumors and their sources in social networks. In: Proceedings of the society of photo-optical instrumentation engineers conference series, Baltimore, pp 1–13
13. Hamidian S, Diab M (2015) Rumor detection and classification for twitter data. In: Proceedings of the 15th international conference on social media technologies, communication, and informatics, Barcelona, pp 71–77
14. Zubiaga A, Liakata M, Procter R (2016) Learning reporting dynamics during breaking news for rumour detection in social media. arXiv preprint arXiv:1610.07363
15. Miller GA (1995) WordNet: a lexical database for English. Commun ACM 38(11):39–41
16. Dong Z, Dong Q (2006) HowNet and the computation of meaning. World Scientific Publishing, Singapore
17. Takahashi T, Igata N (2017) Rumor detection on twitter. In: Proceedings of the 6th international conference on soft computing and intelligent systems, Kobe, pp 452–457
18. Castillo C, Mendoza M, Poblete B (2011) Information credibility on twitter. In: Proceedings of the 20th international conference on world wide web, Hyderabad, pp 675–684
19. Sun S, Liu H, He J (2013) Detecting event rumors on Sina Weibo automatically. In: Proceedings of web technologies and applications, Sydney, pp 120–131
20. Su S, Wan C, Hu Y (2016) Characterizing geographical preferences of Int tourists and the local influential factors in China using geo-tagged photos on social media. Appl Geogr 73(6):26–37
21. Sahin S, Kozat S (2019) Nonuniformly sampled data processing using LSTM networks. IEEE Trans Neural Netw Learn Syst 30(5):1452–1461
22. Ma J, Gao W, Wei Z (2015) Detect rumors using time series of social context information on microblogging websites. In: Proceedings of the 24th ACM international conference on information and knowledge management, Melbourne, pp 1751–1754
23. Shah D, Zaman T (2012) Rumor centrality: a universal source detector. ACM Sigmetrics Perform Eval Rev 40(1):199–210
24. Shah D, Zaman T (2011) Rumors in a network: who's the culprit? IEEE Trans Inf Theory 57(8):5163–5181
25. Pinto PC, Thiran P, Vetterli M (2012) Locating the source of diffusion in large-scale networks. Phys Rev Lett 109(6):068702
26. Xu W, Chen H (2015) Scalable rumor source detection under independent cascade model in online social networks. In: Proceedings of the 2015 11th international conference on mobile ad-hoc and sensor networks, Shenzhen, pp 236–242

Appendix
Mathematical Basis

A.1 Basis of Vectors and Matrices

In depth learning, we often use vectors and matrices. This section briefly introduces the basic knowledge of vectors and matrices.

A.1 1. Vector Basis

(1) Vector

What is a vector? A vector is defined as a quantity with size and direction. In the rectangular coordinate system, given a fixed point a, connect the origin point to point a, form a line from O to a direction, representing the vector, as shown in Fig. A.1.

In the coordinate representation of vectors, $\overrightarrow{OA} = a = (a_1, a_2)$ represents the same vector. The size of the vector is expressed as $|\vec{a}|$. So how do two vectors add up in a rectangular coordinate system? As shown in Fig. A.2: Suppose that the vectors

Fig. A.1. Vector

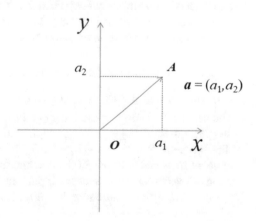

© Tsinghua University Press 2020
X. Liang, *Social Computing with Artificial Intelligence*,
https://doi.org/10.1007/978-981-15-7760-4

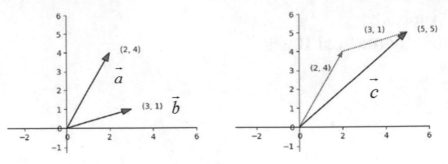

Fig. A.2. Coordinate graph of vector addition

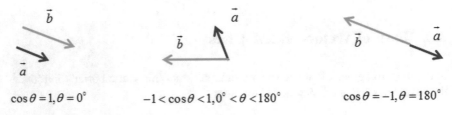

Fig. A.3. Calculation graph of vector inner product

are $\vec{a} = ([\begin{smallmatrix} 2 \\ 4 \end{smallmatrix}])$, $\vec{b} = ([\begin{smallmatrix} 3 \\ 1 \end{smallmatrix}])$, then $\vec{c} = \vec{a} + \vec{b} = \begin{bmatrix} x_1 \\ y_1 \end{bmatrix} + \begin{bmatrix} x_2 \\ y_2 \end{bmatrix} = \begin{bmatrix} x_1 + x_2 \\ y_1 + y_2 \end{bmatrix} = \begin{bmatrix} 5 \\ 5 \end{bmatrix}$.

(2) Inner Product of Vector

The inner product of a vector is a dot product of two vectors with directions. The definition of the formula is $\vec{a}\vec{b} = |\vec{a}||\vec{b}| \cos \theta$, where θ is the angle between the two vectors. Its geometric meaning is used to represent or calculate the angle between two vectors and the projection of vector b in the direction of vector a. It can be seen from the formula that when the two vectors are in the same direction, the inner product gets the maximum value; when the two vectors are in the opposite direction, the inner product gets the minimum value; and when the two vectors are not parallel, the inner product is based on the maximum value and the minimum value (shown in Fig. A.3).

Therefore, the Cauchy Schwartz inequality can be deduced, $-|\vec{a}||\vec{b}| \leq \vec{a} \cdot \vec{b} \leq |\vec{a}||\vec{b}|$. This formula proves that the gradient descent method plays an important role in the next section. How is the inner product of a vector calculated in a coordinate system? Suppose $\vec{a} = [a_1, a_2, ..., a_n]$, $\vec{b} = [b_1, b_2, ..., b_n]$, then the inner product of the two vectors is $\vec{a} \cdot \vec{b} = a_1 b_1 + a_2 b_2 + \cdots + a_n b_n$, which requires the same number of rows and columns of the two one-dimensional vectors. In depth learning, we can use vectors to represent the input value, weight, and so on. Input value $\vec{x} = [x_1, x_2, ..., x_n]$, weight value $\vec{w} = [w_1, w_2, ..., w_n]$.

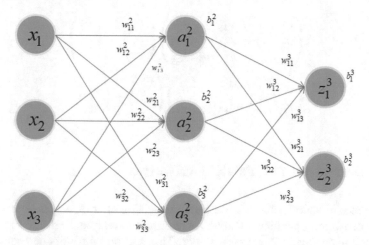

Fig. A.4. Simple neural network structure

A.1 2. Matrix Basis

In deep learning, using matrix to represent neural network will make mathematical formula more concise and clear. Next, let us sort out the matrix knowledge points applied in the neural network.

(1) **Matrix**

Matrix represents a number table, a collection of data, and an array of numbers,

$$A = \begin{bmatrix} a_{11} & a_{12} & a_{13} \\ a_{21} & a_{22} & a_{23} \\ a_{31} & a_{32} & a_{33} \end{bmatrix}.$$

Matrix A is composed of three rows and three columns. The horizontal row is called row and the vertical row is called column. The vector (a_{11}, a_{12}, a_{13}) can be called a row vector in A. The vector $(a_{11}, a_{21}, a_{31})^{\mathrm{T}}$ is a column vector in A. Generally speaking, a matrix can be composed of vectors of the same dimension.

(2) **Matrix representation in neural networks**

Matrix formula is used in computer programming. It is helpful for programming and readers to understand the formula by transforming the relation into matrix form. So how is matrix represented in neural network? As shown in Fig. A.4, it is a simple neural network structure. It includes input layer, hidden layer, and output layer.

It is assumed that the activation function is linear $Z(x) = x$. The matrix is as follows,

$$\begin{pmatrix} a_1^2 \\ a_2^2 \\ a_3^2 \end{pmatrix} = \begin{pmatrix} w_{11}^2 & w_{12}^2 & w_{13}^2 \\ w_{21}^2 & w_{22}^2 & w_{23}^2 \\ w_{31}^2 & w_{32}^2 & w_{33}^2 \end{pmatrix} \begin{pmatrix} x_1 \\ x_2 \\ x_3 \end{pmatrix} + \begin{pmatrix} b_1^2 \\ b_2^2 \\ b_3^2 \end{pmatrix},$$

$$\begin{pmatrix} z_1^3 \\ z_2^3 \end{pmatrix} = \begin{pmatrix} w_{11}^3 & w_{12}^3 & w_{13}^3 \\ w_{21}^3 & w_{22}^3 & w_{23}^3 \end{pmatrix} \begin{pmatrix} a_1^2 \\ a_2^2 \\ a_3^2 \end{pmatrix} + \begin{pmatrix} b_1^3 \\ b_2^3 \end{pmatrix}.$$

A.2   Derivatives and Partial Derivatives

The essence of neural network is the backpropagation of error, which involves some matrix derivation operations. Only by mastering the relevant derivation rules, can we really understand the neural network. This section will introduce some basic knowledge of derivative and partial derivative in detail, hoping to play a role in understanding the knowledge of neural network.

A.2   1. Derivatives in Neural Networks

The derivative function $f'(x)$ of function $y = f(x)$ is defined as,

$$f'(x) = \lim_{\Delta x \to 0} \frac{f(x + \Delta x) - f(x)}{\Delta x}.$$

It can also be expressed as a fraction

$$f'(x) = \frac{dy}{dx}$$

Sigmoid function is one of the most commonly used activation functions in neural networks. In the gradient descent method, the derivative of this function is needed. The sigmoid function is defined as,

$$\sigma(x) = \frac{1}{1 + \exp^{-x}}.$$

The derivation of sigmoid function $\sigma(x)$ is as follows,

$$\sigma'(x) = \left(\frac{1}{1 + \exp^{-x}} \right)' = -\frac{(1 + \exp^{-x})'}{(1 + \exp^{-x})^2} = \frac{\exp^{-x}}{(1 + \exp^{-x})^2}$$

$$= \frac{1 + \exp^{-x} - 1}{(1 + \exp^{-x})^2} = \frac{1}{1 + \exp^{-x}} - \frac{1}{(1 + \exp^{-x})^2}$$

$$= \sigma(x) - \sigma(x)^2.$$

A.2   2. Partial Derivatives in Neural Networks

The derivative of multivariable function is involved in neural network, because we need to treat the weight value and bias value as variables. The derivation method is also applicable to multivariable functions. It should be noted that in multivariable functions, it is necessary to specify which variable to derivative. The derivation of this specified variable is called the partial derivative of this specified variable.

Suppose that the multivariable function $z = f(x, y)$ contains two variables x and y. Let us find the partial derivatives of variables x and y, respectively,

$$\frac{\partial z}{\partial x} = \frac{\partial f(x, y)}{\partial x} = \lim_{\Delta x \to 0} \frac{f(x + \Delta x, y) - f(x, y)}{\Delta x},$$
$$\frac{\partial z}{\partial y} = \frac{\partial f(x, y)}{\partial y} = \lim_{\Delta y \to 0} \frac{f(x, y + \Delta y) - f(x, y)}{\Delta y}.$$

Because the neural network is a multilayer structure, it is necessary to learn the chain rule of derivation of multi variable complex functions, which is necessary for the subsequent error backpropagation.

What is a compound function? Known function $y = g(u)$, $u = f(x)$, this kind of nested function $y = g(f(x))$ is compound function. For the chain rule of multivariable composite function, we know that $z = f(u,v)$, and u and v are functions of x and y, respectively. Then z is also a function of x and y. The partial derivative of z to x and y is

$$\frac{\partial z}{\partial x} = \frac{\partial z}{\partial u}\frac{\partial u}{\partial x} + \frac{\partial z}{\partial v}\frac{\partial v}{\partial x},$$
$$\frac{\partial z}{\partial y} = \frac{\partial z}{\partial u}\frac{\partial u}{\partial y} + \frac{\partial z}{\partial v}\frac{\partial v}{\partial y}$$

A.3   Gradient Descent Method

A.3.1   Derivation Process of One-Dimensional Gradient Descent Method

First of all, what is gradient? Gradient is a vector with size and direction. Generally speaking, gradient means the maximum value of the derivative of a function in the direction of the change point along this direction, that is, the derivative of the function at the current position.

Suppose function $f(x_1, x_2, \ldots, x_n)$, for each point $P(x_1, x_2, \ldots, x_n)$ on the function, we define a vector $\left(\frac{\partial f}{\partial x_1}, \frac{\partial f}{\partial x_2}, \ldots, \frac{\partial f}{\partial x_n} \right)$, this vector is the gradient of function $f(x_1, x_2, \ldots, x_n)$ at point $P(x_1, x_2, \ldots, x_n)$, and let it be $\nabla f(x_1, x_2, \ldots, x_n)$.

The gradient of a unary function $f(x)$ at a certain point is the derivative at that position, which is $f(x)$. The direction of the gradient is the fastest rising direction of the function. The maximum value of the function can be found along the direction of the gradient, and the minimum value of the function can be found along the direction of the minimum gradient. But our cost function needs to find the minimum error, so we should follow the opposite direction of gradient in gradient descent. If $f(x)$ is a convex function, then the gradient descent method can be used to calculate the minimum value of $f(x)$,

$$\nabla = \frac{\mathrm{d}f(x)}{\mathrm{d}x},$$
$$x = x_0 - \eta \cdot \nabla \mathrm{f}(x_0)$$

where x_0 is the starting position, x is the next position after x_0 is updated, η is the learning factor, and $\nabla f(x_0)$ is the gradient of the current position.

First of all, let us understand what is the first-order Taylor series expansion. The first-order Taylor series expansion is

$$f(x) \approx f(x_0) + (x - x_0) \cdot \nabla f(x_0).$$

The following can be obtained by transferring,

$$f(x) - f(x_0) \approx (x - x_0)\nabla f(x_0).$$

The corresponding gradient decline can be expressed as.

$$J(\theta) \approx J(\theta_0) + (\theta - \theta_0) \cdot J'(\theta).$$

According to the concept of local linear approximation of a function, $(\theta - \theta_0)$ is a tiny vector. The closer θ is to θ_0, the better the effect is. So θ and θ_0 can be expressed as follows,

$$\theta - \theta_0 = \lambda d,$$

where the step length λ is scalar, it should be noted that the size of λ can only be tiny. The unit vector of $\theta - \theta_0$ can be expressed as d, so.

$$J(\theta) \approx J(\theta_0) + \lambda d \cdot J'(\theta).$$

Every time we conduct a gradient descent operation, we want the value to be closer to the optimal solution. We want $J(\theta) < J(\theta_0)$, so

$$J(x) - J(x_0) \approx \lambda d \cdot J'(x_0) < 0,$$

$$\lambda d \cdot J'(x_0) < 0,$$

$$d \cdot J'(x_0) < 0.$$

When the inner product points of two vectors multiply, if the two vectors are in the same direction, the inner product gets the maximum value; if the two vectors are in the opposite direction, the inner product gets the minimum value; if the two vectors are not parallel, the inner product is based on the maximum value and the minimum value. In order to minimize the inner product, that is, when two vectors are opposite, to obtain the minimum value is the optimal solution of the loss function,

$$\|d\| \|J'(x_0)\| \cos \theta < 0.$$

When $\cos \theta = -1$, the two vectors are completely opposite, and the minimum value of inner product of vectors is obtained. In the same way, when the direction of unit vector d and gradient $J'(x_0)$ is opposite to each other, the minimum value is obtained, and d is the unit vector, so

$$d = -\frac{J'(x_0)}{J'(x_0)}.$$

When it is taken into $\theta - \theta_0 = \lambda d$, it is obtained that

$$\theta - \theta_0 = \lambda d,$$
$$\theta = \theta_0 + \lambda d,$$
$$\theta = \theta_0 - \lambda \frac{J'(x_0)}{J'(x_0)}.$$

Because $\frac{1}{J'(x_0)}$, and λ are combined into η, the formula of gradient descent method is simplified,

$$\theta = \theta_0 - \eta J'(x_0).$$

It can also be written as,

$$x = x_0 - \eta \nabla f(x_0).$$

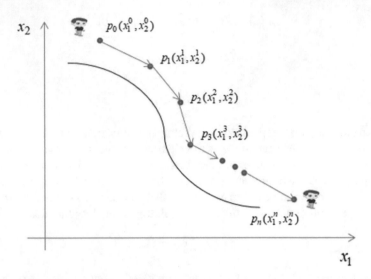

Fig. A.5. Gradient descent of binary function

A.3   2. Gradient Descent Formula of Binary Function

Function $f(x_1, x_2)$ is a binary function of independent variables x_1 and x_2, when x_1 changes Δx_1, x_2 changes Δx_2, the function changes $f(\Delta x_1, \Delta x_2)$. It can be seen from the previous section that when the direction of vector $(\Delta x_1, \Delta x_2)$ and gradient $\left(\frac{\partial f(x_1, x_2)}{\partial x_1}, \frac{\partial f(x_1, x_2)}{\partial x_2} \right)$ is opposite, the function $f(x_1, x_2)$ changes fastest to the minimum value. The gradient descent formula of binary function is.

$$x_1^i = x_1^{i-1} - \eta \frac{\partial f\left(x_1^{i-1}, x_2^{i-1}\right)}{\partial x_1^{i-1}},$$

$$x_2^i = x_2^{i-1} - \eta \frac{\partial f\left(x_1^{i-1}, x_2^{i-1}\right)}{\partial x_2^{i-1}}.$$

where x_1^i represents the next position of variable x_1^{i-1}, x_2^i represents the next position of variable x_2^{i-1}, η is the learning rate, $\frac{\partial f\left(x_1^{i-1}, x_2^{i-1}\right)}{\partial x_1^{i-1}}$ refers to the partial derivative of x_1 at the current position, and $\frac{\partial f\left(x_1^{i-1}, x_2^{i-1}\right)}{\partial x_2^{i-1}}$ refers to the partial derivative of x_2 at the current position. As shown in Fig. A.5.

To find the minimum value of binary function, firstly, the starting position $p_0\left(x_1^0, x_2^0\right)$ is used to reduce the fastest direction by the above formula and moves along this direction with a moving step of η, to $p_1\left(x_1^1, x_2^1\right)$; after moving to $p_1\left(x_1^1, x_2^1\right)$, the moving direction is obtained by the above formula, and the moving step η to $p_2\left(x_1^2, x_2^2\right)$; and so on until the minimum value $p_n\left(x_1^n, x_2^n\right)$, that is, the minimum value of the objective function.

Similarly, binary variable function is extended to multivariate function. When function $f(x_1, x_2, \ldots, x_n)$ is composed of multiple independent variables, the formula of gradient descent method is

$$(\Delta x_1, \Delta x_2, \ldots, \Delta x_n) = -\eta \left(\frac{\partial f}{\partial x_1}, \frac{\partial f}{\partial x_2}, \ldots, \frac{\partial f}{\partial x_n} \right).$$

Like binary variable function, function $f(x_1, x_2, \ldots, x_n)$ moves a small distance η from the starting point $p_0(x_1^0, x_2^0, \ldots, x_n^0)$ in the opposite direction of gradient $\left(\frac{\partial f}{\partial x_1}, \frac{\partial f}{\partial x_2}, \ldots, \frac{\partial f}{\partial x_n} \right)$ to $p_1(x_1^1, x_2^1, \ldots, x_n^1)$, which is the fastest direction of function reduction. Repeat until the minimum value $p_n(x_1^n, x_2^n, \ldots, x_n^n)$ is reached and the optimal solution is get.

Index

© Tsinghua University Press 2020
X. Liang, *Social Computing with Artificial Intelligence*,
https://doi.org/10.1007/978-981-15-7760-4